PUTTING

BIOTECHNOLOGY

TO WORK

BIOPROCESS ENGINEERING

Committee on Bioprocess Engineering

Board on Biology

Commission on Life Sciences

National Research Council

NATIONAL ACADEMY OF SCIENCES
Washington, D.C. 1992

660.6
P 993

This Board on Biology study was supported by the National Science Foundation, the Department of Energy, the U. S. Department of Agriculture, the National Aeronautics and Space Administration, the National Institute of Standards and Technology, and the National Academy of Engineering.

Library of Congress Catalog Card Number 92-61717

International Standard Book Number 0-309-04785-4

Additional copies of this report are available from:
National Academy Press
2101 Constitution Avenue, N.W.
Washington, D.C. 20418

B-019

COMMITTEE ON BIOPROCESS ENGINEERING

Michael R. Ladisch (Chairman), Purdue University, West Lafayette,
Indiana
Charles L. Cooney, Massachusetts Institute of Technology, Cambridge,
Massachusetts
Robert C. Dean, Jr., Dean Technology, Inc., Lebanon, New Hampshire
Arthur E. Humphrey, Pennsylvania State University, State College,
Pennsylvania
T. Kent Kirk, U.S. Department of Agriculture, Madison, Wisconsin
Larry V. McIntire, Rice University, Houston, Texas
Alan S. Michaels, Alan Sherman Michaels, Sc.D., Inc., Chestnut Hill,
Massachusetts
Paula Myers-Keith, Pitman-Moore, Inc., Terre Haute, Indiana
Dewey D.Y. Ryu, University of California, Davis, California
James R. Swartz, Genentech, Inc., San Francisco, California
Daniel I.C. Wang, Massachusetts Institute of Technology, Cambridge,
Massachusetts
Janet Westpheling, University of Georgia, Athens, Georgia
George M. Whitesides, Harvard University, Cambridge, Massachusetts

National Research Council Staff:

John E. Burris, Study Director
Oskar Zaborsky, Study Director (until April 1992)
Marietta Toal, Administrative Secretary
Norman Grossblatt, Editor

The National Academy of Sciences is a private, nonprofit, self-perpetuating society of distinguished scholars engaged in scientific and engineering research, dedicated to the furtherance of science and technology and to their use for the general welfare. Upon the authority of the charter granted to it by the Congress in 1863, the Academy has a mandate that requires it to advise the federal government on scientific and technical matters. Dr. Frank Press is president of the National Academy of Sciences.

The National Academy of Engineering was established in 1964, under the charter of the National Academy of Sciences, as a parallel organization of outstanding engineers. It is autonomous in its administration and in the selection of its members, sharing with the National Academy of Sciences the responsibility for advising the federal government. The National Academy of Engineering also sponsors engineering programs aimed at meeting national needs, encourages education and research, and recognizes the superior achievements of engineers. Dr. Robert M. White is president of the National Academy of Engineering.

The Institute of Medicine was established in 1970 by the National Academy of Sciences to secure the services of eminent members of appropriate professions in the examination of policy matters pertaining to the health of the public. The Institute acts under the responsibility given to the National Academy of Sciences by its congressional charter to be an advisor to the federal government and, upon its own initiative, to identify issues of medical care, research, and education. Dr. Kenneth Shine is president of the Institute of Medicine.

The National Research Council was organized by the National Academy of Sciences in 1916 to associate the broad community of science and technology with the Academy's purposes of furthering knowledge and of advising the federal government. Functioning in accordance with general policies determined by the Academy, the Council has become the principal operating agency of both the National Academy of Sciences and the National Academy of Engineering in providing services to the government, the public, and the scientific and engineering communities. The Council is administered jointly by both Academies and the Institute of Medicine. Dr. Frank Press and Dr. Robert M. White are chairman and vice chairman, respectively, of the National Research Council.

Preface

Biotechnology is broadly defined in a 1991 Office of Technology Assessment report as "any technique that uses living organisms (or parts of organisms) to make or modify products, to improve plants or animals, or to develop microorganisms for specific uses." This technology has been instrumental in the development and implementation of processes for the manufacture of antibiotics and other pharmaceuticals, industrial sugars, alcohols, amino acids and other organic acids, foods, and specialty products through the application of microbiology, fermentation, enzymes, and separation technology. Engineers, working with life scientists, often achieved scale-up to industrial production in remarkably short periods. A relatively small number helped to catalyze, over a period of 50 years, the growth of the pharmaceutical, food, agricultural-processing, and specialty-product sectors of the U.S. economy to the point where sales now exceed $100 billion/year.

The introduction of the new biotechnology since 1970 enabled directed manipulation of the cell's genetic machinery through recombinant-DNA techniques and cell fusion. Its application on an industrial scale since 1979 has fundamentally expanded the utility of biological systems and positioned a number of industries for explosive global growth. Scientists and engineers can now change the genetic makeup of microbial, plant, and animal cells to confer new characteristics. Biological molecules, for which there is no other means of industrial production, can now be generated. Existing industrial organisms can be systematically altered (i.e., engineered) to enhance their function and to produce useful products in new ways. The new biotechnology, combined with the existing industrial, government, and university infrastructure in biotechnology and the pervasive influence of biological substances in everyday life, has set the stage for unprecedented growth in products, markets, and expectations.

Substantial manufacturing capability will be needed to bring about the full application of biotechnology for the benefit of society. A wide array of engineering fundamentals applied to biological systems will be required to produce and purify biological products on a commercial scale. Bioprocess engineers will be essential for translating the discoveries of biotechnology into tangible commercial products, thereby putting biotechnology to work. The Committee on Bioprocess Engineering was convened in the National Research Council's Board on Biology to address issues that are of critical importance if the nation is to reap the full benefits of its success in fundamental biotechnology research: What discoveries and concepts in biology and chemistry are important to bioprocess engineering? What barriers to their exploitation exist? What is the position of the United States in relation to other countries' efforts in bioprocess engineering, especially those of Japan and Germany? What actions are required to ensure that research and training are adequately organized and supported so that the United States can maintain and improve its position? The committee met five times from May 1991 to May 1992 and found that much needs to be done, and done quickly. This report represents a consensus of the committee, which hopes to impart a sense of urgency to the planning for bioprocess-engineering needs in biotechnology manufacturing in the United States.

The committee carefully considered the best way to present its findings and to organize the report, given the wide range of products, services, and needs that will be affected by bioprocess engineering and the diverse backgrounds of those who will read this report. The committee decided that the reader should first be provided a definition of bioprocess engineering, a discussion of its economic impact on biotechnology, and a summary of major barriers to the exploitation of biotechnology. Further definitions and a historical perspective were to be addressed in Chapter 2, to illustrate the role of bioprocess engineering in a substantial portion of the economic sectors of the United States. We decided to present the current status of U.S. capabilities, and those of Japan and Europe next, because of the importance of international competitiveness for U.S. economic activities, particularly those affected by bioprocessing. The many areas already affected by bioprocess engineering are presented in Chapter 4, to help the reader become more aware of the language and technologies encompassed by biotechnology. Having "set the stage," the committee chose to present, in Chapter 5, what needs to be done now to address needs that will not be fully understood for some years to come. Chapter 6 addresses future scenarios of biotechnology development and how the education, training, research, and technology-transfer issues related to current opportunities (described in Chapter 5) will prepare bioprocessing to address future needs.

The committee thanks those who contributed to its work and shared their expertise at our meetings. In particular, we would like to thank the National Science Foundation (NSF) Directorate for Engineering (Biotechnology Program and Divisions of Biological and Critical Systems, Engineering Education Centers, and Chemical and Thermal Systems), the NSF Directorate for Biological Sciences (Divisions of Behavioral and Cognitive Sciences, Biological Instrumentation and Resources, Social and Economic Science, and Molecular Biosciences), the Department of Energy (offices of Industrial Technologies, Fossil Energy, and Alcohol Fuels), the Department of Agriculture (Office of the Assistant Secretary for Science and Education and Agricultural Research Service), the National Aeronautics and Space Administration (Life Sciences Division, Microgravity Science Division, and Office of Commercial Programs), the Department of Commerce (National Institute of Standards and Technology, Chemical Science and Technology Laboratory), and the National Academy of Engineering for funding this study. Duane Bruley, Luther Williams, and Carl Hall of NSF deserve special mention for their support of this study in its early stages and Fred Heineken of NSF for serving as the contact person of the lead agency on logistic matters.

The chairman thanks Purdue University for making time available to carry out the tasks associated with the committee's work, Carolyn Wasson for excellent assistance in preparing the various drafts of this report, and Norma Leuck for coordinating the numerous communications with committee members. The chairman also thanks Michael Shuler of Cornell University for making his expertise available and contributing to the technical completeness of the report; Charles Scott of Oak Ridge National Laboratories for comments on bioprocess-engineering needs; Edith Munroe of the Corn Refiners Association, Inc., and Matthew Rendlemen and Betsy Kuhn of the Department of Agriculture Economic Research Service for helpful information on value-added products from corn; and Karl H. Kroner of the German National Research Center for Biotechnology (GBF) for providing information on German bioprocess engineering. The committee thanks Donald Henninger, Doug Ming, and Glenn Spaulding of the National Aeronautics and Space Administration Johnson Space Center for arranging a subcommittee visit to the center and is indebted to Marietta Toal of the Board on Biology for her excellent assistance with committee meetings. The committee also thanks numerous individuals and organizations that rapidly responded to inquiries from the committee. Norman Grossblatt, of the National Research Council's Commission on Life Sciences, edited the report. Special thanks are due to Oskar Zaborsky, director of the Board on Biology, whose vision, hard work, and many capabilities enabled this study to be initiated and carried out in a timely manner, and to him and John Burris, executive director of the Commission on Life Sciences, for the long hours they spent in guiding this report through many drafts to its successful conclusion.

Michael R. Ladisch, Chairman
Committee on Bioprocess Engineering

Contents

xi

Executive Summary

The United States has dominated the discovery phase of biology and laid the groundwork for commercialization of biotechnology. Biotechnology-derived products already affect human health, nutrition, and environmental improvement and will grow to provide new products and employment in new industries. Worldwide markets for biotechnology-derived products are projected to grow to at least $50 billion per year within the next 10 years, and our global trading partners are concentrating their resources on translating the discoveries of biology into economically viable products through bioprocess engineering.

Bioprocess engineering is the subdiscipline within biotechnology that is responsible for translating life-science discoveries into practical products, processes, or systems capable of serving the needs of society. It is critical in moving newly discovered bioproducts into the hands of the consuming public. Although the United States has nurtured the discovery phase of biotechnology, it has not been aggressive in developing bioprocess engineering.

BIOPROCESS ENGINEERING AND GLOBAL COMPETITIVENESS

The importance of engineering capability in achieving and maintaining global competitiveness is compelling; witness the growth of the pharmaceutical industry after the development of penicillin production during World War II and of the computer and electronics industry after the discovery of the transistor. The strength of the United States in engineering and manufacturing technology made major contributions to America's early dominance of world markets in both instances.

1

The U.S. ambivalence toward bioprocess engineering is an inadvertent consequence of the high biochemical potency of the protein-based pharmaceuticals introduced between 1982 and 1989 whose worldwide markets are measured in kilograms per year and whose sales are in billions of dollars. But the situation is changing. The emerging families of food, agricultural products, and industrial chemicals to be generated by biological routes, as well as the biopharmaceutical products now in development, will have markets measured in thousands of kilograms, or more, and will require innovative manufacturing techniques.

The participation of the United States in the expanding bioproducts markets will necessitate world-class bioprocess engineering. Comparison of the global competitive position of the United States with that of other technologically advanced nations in biotechnology and bioprocess engineering reveals that

• The United States continues to be the world leader in basic health-science and life-science elements of biotechnology.
• Japan leads in applied microbiology and biocatalysis and is effectively coordinating government, industrial, and academic resources in biotechnology and bioprocess-engineering development.
• Europe matches Japan in progress in applied biocatalysis and is establishing a strong, government-supported technology-transfer infrastructure between industry and academe with emphasis on bioprocess engineering.

World competition in biotechnology and other industries that depend on bioprocess engineering will be keen because of the notable capabilities in and commitments to biologically relevant manufacturing and bioprocess development among industrially developed nations. It is debatable whether the United States can be dominant (or even competitive) in bioprocessing: university research and training programs are projected to grow by 75% in the best case while the industry grows by 1,000% in the next 10 years. The committee concurs with the Federal Coordinating Council on Science, Engineering, and Technology assessment that "manufacturing/bioprocessing is an area in which biotechnology offers vast potential rewards. The total federal investment of $99 million in FY 1992 is small in proportion to its potential."

The committee recommends that the U.S. government promptly take action and provide suitable incentives to establish a national program in bioprocess-engineering research, development, education, and technology transfer. That will require that the existing resources of government, industry, and academe collaborate in

• Rapidly translating scientific discoveries into marketable products and processes.

• Promoting cross-disciplinary research and education and thereby fostering innovative, multidisciplinary solutions to important bioprocessing problems.

• Providing a growing cadre of bioprocess engineers to meet the needs of an expanding bioprocess industry.

OPPORTUNITIES

The committee addressed trends in biotechnology that are likely to have important worldwide social and financial impact within the next 10 years. In this context, current commercial activities related to biotechnology and biotechnology products are dominated by biopharmaceutical biologics, such as insulin, tissue plasminogen activator, and erythropoietin. Innovative bioprocess engineering in the manufacture of these products can lead to improvements in product recovery, product purity, process safety, and reduced manufacturing and quality-control costs. The need for such process innovation will intensify as patent protection for these products expires, global competition for international markets increases, and regulatory procedures that would otherwise slow introduction of new bioprocess technologies are streamlined. Health-care products emerging from biotechnology will be consumed in much larger quantities around the world than they are now (examples include recombinant hemoglobin, recombinant albumin, and conjugate vaccines). These second-generation products will require large-scale manufacturing facilities that handle biological systems; and bioprocess engineering will be a *sine qua non* for successful commercialization of the products.

Bioprocess engineers will be employed in applying the new biology to producing smaller molecules and specialty bioproducts. These are in a category where the challenge is to apply bioprocessing to obtain value-added products and to engineer large-scale, integrated processes that use agricultural and forestry-based materials and other renewable resources. Bioproducts for use in food production and in foods (animal health-care biologics, biological plant-growth promoters and pesticides, nutritional supplements, and food additives) present large-tonnage product opportunities that can be tapped in the coming decade, provided that suitably efficient and economical manufacturing facilities can be designed and built. Such capabilities do not exist, and their creation is a major challenge for bioprocess engineering. The use of biomass for the production of industrial chemicals and of liquid and gaseous fuels represents a major hope for reducing U.S. dependence on imported hydrocarbons. The processing of renewable resources must have high national priority in the coming decade, so that the necessary know-how and production infrastructure for its practical implementation can be developed. Bioprocessing in space presents unique opportunities, particularly in bioregenerative life support and as a research platform for the study of new types of manufacturing processes.

Bioprocessing for protection and beneficiation of the environment represents another large and important opportunity. Biological processes could offer alternatives to environmentally polluting or fossil-fuel-consuming manufacturing processes and could help to remove toxic pollutants from industrial and municipal wastes. Bioremediation's promise is in its potentially lower cost, compared with other types of technology for cleaning up the environment.

NEEDS

Generic applied research is critical to the optimal exploitation of bioprocess engineering by industry, in that it addresses technologies that are too risky for companies or that require too long a period for results. This category of research bridges the gap between basic biological science that is carried out by university and government laboratories and the industrial applied research that assists in converting biotechnology into products and services. For biopharmaceuticals, needs identified by the committee are to

- Improve analytical methods that facilitate rapid testing of products for purity and activity.
- Develop high-resolution protein-purification methods for scaleup and application in the industrial manufacture of ultrapure products.
- Develop process-control technology for integrating biological production sequences into stable and robust automated manufacturing systems.
- Enhance biological and biochemical technology for increasing the efficiency of protein folding and improving the expression of recombinant proteins.

For specialty bioproducts and industrial chemicals, key needs are to

- Develop separation and purification technologies that are specially adapted to the recovery of products from dilute aqueous streams characteristic of materials derived from microbial fermentation, plant cell culture, or whole plant material.
- Develop processing technologies that will facilitate the economical conversion of cellulose-based materials into industrial chemicals and fuels.
- Develop specially adapted or genetically altered microorganisms that can transform biomass materials into industrial chemicals and other products.
- Develop bioproduct manufacturing processes that are controlled and regulated and have predictable performance.

Appropriate bioreactor design and operating conditions must be implemented on scaleup to ensure that product characteristics are maintained, regardless of the type of product. Bioprocess engineers are particularly well suited to integrate bioreaction engineering concepts with the subtleties of cellular metabolism to achieve the necessary product qualities.

Bioprocess-engineering input is important for environmental applications of biotechnology, where the needs are to

• Study the role of microbial interactions in degrading of toxic wastes in the environment and detoxifying industrial wastes at the plant site.
• Define standards by which the effects of bioprocessing in detoxifying wastes will be measured.
• Implement bioprocess-engineering methods in the design of waste-processing technologies.

RECOMMENDATIONS

To meet the global challenges of competition in industrialization of bio-technology and to address national needs, the committee recommends

• A coordinated, long-term plan of research, development, training, and education in bioprocess engineering, with well-defined goals that involve participation of industry, academe, and the federal government.
• A research and educational program in bioprocess engineering that emphasizes cross-disciplinary interactions between scientists and engineers and a multidisciplinary team approach to problem-solving, which has historically been the keystone of success in American industrial development.
• Increased cooperation between industry and the Food and Drug Administration for the express purpose of developing quality-control methods and standards and good manufacturing practices for the manufacture of biotechnology products.

Sustained funding by the federal government is essential to the success of research and education programs for training bioprocess engineers, as is the participation of industry—in planning, training, and supply of physical and financial resources.

The ability of the United States to sustain a dominant global position in biotechnology lies in developing a strong resource base for bioprocess engineering and bioproduct manufacturing and maintaining its primacy in basic life-science research. The United States has made an enormous, and enormously successful, investment in basic biological science. To protect the investment and to capitalize on it, there must now also be an investment in bioprocess engineering.

A PLAN FOR ACTION

The discoveries emanating from the basic life sciences provide the knowledge that supports new concepts for biologically based products and manufacturing systems. The committee strongly recommends that federal funding of research in biotechnology be extended to support efforts that provide the science and technology base for producing and manufacturing products from biology. Targeted long-term research support would speed the development of commercial products, provide the trained personnel needed to support industrial activities, protect entry-level U.S. products, provide the basis for low-cost production of the largest-volume (and highest-revenue) products, and help to integrate processes and concepts from biological science and bioprocess engineering.

The Committee on Bioprocess Engineering recommends these actions to improve U.S. strength in bioprocess engineering and U.S. competitiveness in commercial biotechnology.

Human-Resources Development

The committee recommends a major commitment to developing the human-resources base through funding of research programs in universities, continuing-education programs, and research directed toward industrial problems (applied-engineering research) by the cognizant government agencies, including the National Science Foundation (NSF), the National Institutes of Health (NIH), the Department of Energy (DOE), and the Department of Agriculture (USDA). New resources must be provided to strengthen the infrastructure for bioprocess engineering and biotechnology in this context. A major commitment is needed to educate personnel skilled in bioprocess engineering. These are the individuals who will develop bioprocesses and support biologically based manufacturing technologies if the U.S. biotechnology industry is to remain competitive with those of Japan and Europe.

Assessing Developments Abroad

The committee recommends vigorous efforts in technology assessment in Japan and Europe and support for exchange-scholar, exchange-student, and collaborative research programs. For bioprocess engineering, particular emphasis should be placed on tracking developments in process technology for manufacture of new bioproducts.

Germany, Switzerland, Austria, the United Kingdom, France, Scandinavia, Italy, the Netherlands, and other European nations have a strong base serving biotechnology in products and services. Given the upcoming economic unification of Europe, we recommend a separate study on bioprocessing in Europe.

Cross-disciplinary Research

Several government-agency programs, including those of the NSF Engineering Research Center Initiative and the NIH Interdisciplinary Biotechnology Training Grant Program in the National Institute of General Medical Sciences, foster cross-disciplinary and interdisciplinary training. Cross-disciplinary research should be part of the training of the bioprocess engineer and include activities at the postgraduate level. For example, a postdoctoral scholarship program for biological scientists to gain exposure to engineering activities should be considered. **The committee recommends that cross-disciplinary interactions continue to be fostered by the programs of NSF, NIH, USDA, and DOE.**

Promoting Awareness of Importance of Manufacturing Technology

The committee strongly recommends formulating a federal strategy for fostering increased awareness of the importance of manufacturing technology in the research and university communities through education and training. Postdoctoral and graduate students should have contact with issues in manufacturing through research, course work, teaching laboratories, and industrial experiences. Continuing education is also critical for bioprocess engineering because of the rapidity of advances in the biological sciences and should be part of the training offered by universities to leaders in the bioproduct industry. Such programs should be created by industry, universities, and government in a cooperative fashion. Teaching laboratories for bioprocess engineering should be upgraded so that they can provide a high-quality training experience for a larger number of students.

Competitive-Grants Program

The committee recommends that research funding be allocated to topics listed in this report through a competitive-grants program for bioprocesses in the manufacture of biopharmaceuticals and other bioproducts that cover a wide range of biological, chemical, and engineering disciplines. We believe that structuring the research in a manner that requires an industry-university or industry-government interaction would catalyze further research.

Role of National Laboratories and Research Centers

Industry involvement and the special facilities and capabilities of government laboratories, such as those under the auspices of USDA and DOE, could help to speed adaptation of some types of new bioprocesses on a commercial scale. Similarly, the NSF Engineering Research Centers program and more recently the National Aeronautics and Space Administration Scientific Centers for research and training provide cross-disciplinary environments for research related to manufacturing or large-scale systems. **The committee recommends that these laboratories and centers be examined as models and applied, in a suitably modified form, to the processing of renewable resources.**

Bioprocessing for Cleanup of Environmental Hazards

The committee recommends an analysis of the costs of biological treatment compared with other technologies. Bioremediation promises lower costs than other types of technology for cleaning up certain environmental hazards.

1

Introduction

1.1 BIOTECHNOLOGY: THE CHALLENGE OF THE TWENTY-FIRST CENTURY

Biotechnology, which covers a broad segment of science and its industrial and societal applications, has commanded worldwide attention over the last decade because of its perceived potential impact on the quality of life. In simple terms, biotechnology is the application of science and engineering to the use of living organisms or substances derived from them, to generate products or to perform functions that can benefit the human condition. The products include substances that can help to diagnose, prevent, or cure diseases of humans and animals; to enhance the productivity of or eliminate pests that affect crops; or to replace chemicals or other materials that consume nonrenewable resources or create environmental hazards. The functions include the purification of water and air and the generation of energy or industrial chemicals with minimal environmental impact.

1.2 WHAT IS BIOPROCESS ENGINEERING?

Bioprocess engineering is the subdiscipline within biotechnology that is responsible for translating the discoveries of life science into practical products, processes, or systems that can serve the needs of society. The bioprocess engineer has many missions. Although the most visible today is the production of biopharmaceuticals, bioprocess engineering also has a major role in the existing multibillion-dollar fermentation industries responsible for the production of ethanol, amino acids and other organic acids, antibiotics, and other specialty products.

1.3 THE COMMITTEE'S APPROACH

This committee was convened to address the question of whether the nation's present and future capabilities in bioprocess engineering are commensurate with current and emerging global demands for products and services that are evolving from modern biological science. The committee has elected to focus its attention on the trends in biotechnology that are likely to have important worldwide social and financial impact within the next decade. As is elaborated in this report, the committee's consensus is that America's human and technological resources in bioprocess engineering should be increased to meet the growing demands of biotechnology within the coming decade. The committee suggests a national strategy for establishing and maintaining America's global stature in this important field.

1.4 ECONOMIC IMPACT OF BIOTECHNOLOGY

The justification for commercialization of the fruits of any scientific endeavor is the potential for providing marketable goods and services and thereby generating gainful employment and return on invested capital. Basic discoveries in life science within the last 10 years have already created a family of novel biopharmaceutical products with new therapeutic and prophylactic potential. Worldwide annual sales have grown from zero in 1980 to $4 billion in 1991. Further developments in biopharmaceuticals alone are predicted to lead to expansion in global annual sales of $30-50 billion by the year 2000. In addition, recent developments in biological waste treatment and environmental bioremediation are projected to create new industries in waste treatment and modification of chemical processes for waste minimization. Beyond estimation are the enormous savings to be realized after the new technology is implemented. Other growth opportunities for biotechnology are in agricultural chemicals, foods and nutritional supplements, specialty and commodity chemicals, and liquid and gaseous fuels derived from biomass. In many respects, the predicted growth of biotechnology-based industry resembles the immense growth of the pharmaceutical industry after the discovery of penicillin and of the electronics and computer industries after the discovery of the transistor.

The historical record over the last half-century has taught us that, as every new technology matures and reveals large potential markets for its products, price and quality become the dominant factors in market competition; skill and ingenuity in manufacturing are determinants of competitive position and of product price and quality. One of the committee's primary concerns is that the biotechnology industry, as it matures, will suffer the same competitive ills that have beset so many core industries in America.

America has dominated the discovery phase of biotechnology for the last 2 decades, but it has failed to acknowledge the importance of bioprocess engineering in capitalizing on its discoveries in world markets. Our global trading partners (in particular, Japan and Germany, but with the evolving European Economic Community as an even more imposing challenger) accept America's leadership in basic biological science and have yielded to America as the principal source of basic information in biotechnology. Those countries are concentrating their intellectual and fiscal resources on translating that information into industrial practice while maintaining a strong effort in basic research. Consortia of leaders in government, industry, and academe in those countries are devising and implementing strategies for reducing basic scientific discoveries to practice.

The economic potential of biotechnology is not perceived solely by the United States; it is recognized as an opportunity by other technologically advanced countries. Exploitation of this opportunity will require national commitments, dedication of resources, and organization for action. This report attempts to propose a strategy by which the nation can meet this challenge.

1.5 ROLE OF BIOPROCESS ENGINEERING

The role of bioprocess engineering in the successful commercialization of biotechnology is not fully understood by our national government, industrial, and academic leadership. That is in large measure because first-generation biopharmaceutical products have been successfully produced with only secondary concern for costs of manufacturing. However, products now under development will require novel techniques and more efficient and economical processes. Hence, our participation in the expanding bioproducts market will necessitate an expanded role of bioprocess engineering. This is all the more important because bioprocess engineering could have a profound effect on the existing fermentation industry.

1.6 BARRIERS TO EXPLOITATION OF BIOTECHNOLOGY

A long-range program to strengthen America's base in bioprocess-engineering research, development, and education will contribute to its commercialization of biotechnology. Conduct of the program will require the close collaboration and combined resources of the federal government, industry, and academe.

The economic and social culture of the United States is reflected in a number of barriers:

• Legislative barriers and competitive forces in our free-enterprise economy obstruct cooperation between companies within an industry as being restrictive of free trade.

• The investment and financial community is reluctant to provide financing for enterprises with long payout times.

• Regulatory agencies are unable to evaluate new products and processes rapidly.

• Bioprocess engineering is by its very nature an interdisciplinary endeavor, and the culture of the American university is still evolving a coherent approach to cross-disciplinary research and education.

• Academic traditions place high priority on fundamental research in graduate education. Current needs in bioprocess-engineering research are for the application of established engineering principles. A balanced approach is now needed to exploit opportunities in biotechnology.

It is the committee's consensus that those challenges must be met and suitable incentives must be provided, if a productive national program in bioprocess-engineering research, development, and education is to be created and implemented.

2

The Challenge

The biotechnology industry has been successful in translating basic research in the biological sciences and molecular biology into very high-value-added products. Particular emphasis has been on biopharmaceuticals for the treatment of such catastrophic illnesses as cancer, heart disease, and kidney diseases. A second generation of bioproducts is now being developed whose price-cost difference will be much lower: intermediate-value biopharmaceutical and pharmaceutical products, specialty chemicals, and materials and chemicals derived from renewable resources. Bioprocess engineering is critical for the economical development of the new products, particularly as the difference between price and cost decreases and profitability becomes an important function of production costs.

2.1 TRANSLATING SCIENCE INTO PRODUCTS

"Science and the application of science . . . are linked as the fruit is to the tree." Louis Pasteur

Engineering innovation and the development of enabling technologies are the essence of translating science into products. The first step of innovation is discovery of a phenomenon, such as the efficacy of a *Penicillium* culture against *Staphylococcus aureus* (1928), the ability of xylose isomerase to catalyze formation of fructose from glucose (1957), or the ability of enzymes (known as Type II restriction endonucleases) to cleave specific sites in DNA (1970). Recombinant-DNA technology enables development and manufacture of products that would otherwise not be possible.

The next essential element of engineering innovation is an economic opportunity arising from a potential societal benefit. The opportunity can be made more attractive by an emergency or economic perturbation that produces unique market conditions, usually for a short period. For instance, World War II provided the incentive in the case of penicillin; high sugar prices and an abundance of corn in 1965-1980 promoted development of the U.S. high-fructose corn syrup industry based on glucose isomerase; and an anticipated decline in U.S. meat consumption, and therefore in the availability of animal pancreas, might have prompted Eli Lilly Co. to convene a meeting in 1976 on applying recombinant-DNA methods to produce insulin.

The jump from science to technology involves the convergence of preparation with opportunity, when market forces (and in some cases government policy) present economic incentives for introducing a new product. In the case of penicillin, World War II prompted government to promote industrial research on antibiotic production and other subjects through taxation policies. High-fructose corn syrup production was encouraged by a government-defined minimal sugar price. Recombinant insulin was developed rapidly to meet the needs of millions of Americans. In all three cases, a research infrastructure, a fundamental knowledge of the product, and technically capable scientists and engineers in universities, industry, and national government laboratories were in place when the opportunity arose.

The economic impacts have been tremendous. It can be argued that the development of penicillin led to the current U.S. pharmaceutical industry in which the top eight companies had sales of $31.6 billion in 1991. However, 15 years passed between discovery and application, and another 10 years before antibiotic production became an established large-scale endeavor. The lag between discovery of glucose isomerase and the first process was 10 years. Glucose isomerase and, later, application of very-large-scale liquid chromatography resulted in the growth of enzymatically produced 42%-55% high-fructose corn syrups from none in 1966 to 12.7 billion pounds (dry basis) in the United States in 1991 (Antrim et al., 1979; Buzzanell et al., 1992). Recombinant insulin came on the market 12 years after the discovery of Type II restriction endonucleases in 1970; this rapidity reflects not only the technical prowess of the companies involved, but also their ability to deal effectively with the federal government's regulations.

Today there are 16 approved biotechnology-produced drugs and vaccines, which would not exist in the absence of molecular biology. Another 120 are in various stages of federal review and approval (Burrill and Lee, 1991). The developments in biopharmaceuticals suggest a clear need for long-term commitment to preparing scientists and engineers to deal with translating science into production. The need is particularly clear in the light of the projected 10- to 20-fold growth of the industry in the next 10 years, developments in bioprocessing of renewable resources, the desire to apply biotechnology to environmental issues, and the nation's goal of a long-term human presence in space. Meeting that commitment will require a strong program for research and education related to bioprocess engineering.

Vignette 1 illustrates the key elements of applying bioprocess engineering for "putting biology into the bottle." Tremendous screening efforts coupled with a mechanistic understanding of disease-causing processes are needed to identify new therapeutic agents. After discovery of a new substance, enough of it must be obtained for testing with animals and later, if

---

Vignette 1

**The Scaleup of Penicillin Production—
Parallels to the New Biotechnology?**

Alexander Fleming showed that *Staphylococcus* cultures were inhibited by growing colonies of *Penicillium notatum* in 1928 and then identified the antimicrobial activity as due to a secreted substance, i.e., penicillin. Florey and Chain rediscovered penicillin in 1939 and found it to be most effective in treating infections. However, the supply was limited, chemical synthesis proved to be difficult, and the war complicated developmental efforts in England. Scaleup to obtain larger amounts was initiated by a joint Anglo-American effort and included the involvement of Merck, Pfizer, Squibb, and U.S. government laboratories.

The demand for penicillin in 1941-1943 exceeded the amount that could be produced by surface culturing techniques, even when higher-yielding strains of *P. notatum* and *P. chrysogenum* were used. The microorganisms were first grown on the surface of moist bran in milk-bottle-size vessels having a volume of 1-2 L (Shuler and Kargi, 1991). The need for a more efficient manufacturing approach quickly became apparent. The solution to scaleup evolved from a combination of applied microbiology and fermentation engineering.

The discovery, on a moldy cantaloupe in Peoria, Ill., of a strain of *P. chrysogenum* that could be grown in a submerged fermentation was an enabling factor. Scaleup from 1 L to 100,000 L followed (Aiba et al., 1973). Numerous engineering challenges needed to be addressed, from maintaining growing conditions that excluded contaminating organisms to aerating large volumes of fermentation broth. The penicillin itself was labile and thus required development of appropriate purification procedures. As a result of cooperative efforts between government and industry, full-scale production was quickly achieved with microbiologists and chemical engineers working together—i.e., the first bioprocess-engineering teams. Government tax policy further encouraged major industrial involvement in what was considered, at the time, to be the very risky technology of pharmaceutical manufacture by large-scale, submerged, aerated fermentation.

Penicillin productivity increased from about 0.001 g/L in 1941 to over 50 g/L by 1970; that led to a decrease in cost by a factor of more than 1,000 and the classification of penicillin as a bulk material. In the meantime, different types of antibiotics, effective against a range of diseases, were developed; 5,500 were identified between 1945 and 1981, of which about 100 reached the market (Hacking, 1986).

warranted, for treating humans. Bioprocess engineering is critical in rapidly scaling up production of a promising new therapeutic agent so that sufficient quantities of an ultrapure substance are available for testing. If the trials are successful, manufacturing capability must be developed quickly to make the new drug widely available.

Over 100 biopharmaceutical products are in various stages of clinical testing (OTA, 1991). That is almost as many antibiotics as have been brought to market within the last 50 years. The number illustrates the acceleration of pharmaceutical development and the coming need for increased numbers of engineers who have cross-disciplinary training in biology and engineering to scale up the manufacture of new classes of therapeutic compounds.

2.2 BIOPHARMACEUTICALS

Total global sales of all pharmaceuticals are about $150 billion a year, of which about one-third is in the United States (Abelson, 1992). The pharmaceutical products of the new rDNA and hybridoma technology currently make up a small but rapidly growing fraction of the total (estimated U.S. sales, $4 billion in 1991). By the end of this century, it is estimated that total U.S. sales of these new products will exceed $30 billion a year, dominated by therapeutic proteins. This growth is anticipated also to include chemotherapeutic compounds, polysaccharides, vaccines, and diagnostics, as well as therapeutic proteins. This exciting potential has spawned hundreds of new biotechnology companies in the United States. In June 1991, the market capitalization of the industry was $35 billion (Burrill and Lee, 1991). Clearly an important new industry has been born.

However, the recent purchase of 60% of Genentech, Inc., by Roche Holdings, a Swiss company, and the intensive activity in the Japanese pharmaceutical industry suggest that Americans cannot take for granted our current lead in this important and expanding field. The goal of this section is to assess the role of bioprocess engineering in determining competitive position in the biopharmaceutical industry.

2.2.1 New Technology Required for Biopharmaceuticals

The new industry has, to a large extent, required fundamentally new bioprocess technology. A few protein pharmaceuticals had previously been marketed, but rDNA and hybridoma technology has brought such dramatic changes to the manner of production and the range of possible products that a substantially new technology base has had to be developed (see Vignette 2). One characteristic of the new technology is that it enables the design and optimization of the production organism to an unprecedented degree. Another is that the technology must apply to a class of products that, al-

Vignette 2

Bioprocess Engineering for Early Biopharmaceuticals

engineering: the application of science and mathematics by which the properties of matter and the sources of energy in nature are made useful to people in
structures, machines, products, systems, and processes. [From *Webster's Ninth New
Collegiate Dictionary*, Springfield, Mass.: Merriam-Webster Inc., 1984]

The tremendous power of rDNA and hybridoma technology has dramatically
expanded our capabilities for the development of protein pharmaceuticals. At the
same time, it has blurred the boundaries between those who pursue basic biological
knowledge and those who apply that knowledge to bring beneficial products and
services to society. The definition of "engineering" shown above now describes the
design and construction of a new living organism just as well as it describes the
design and construction of the bioreactor used to cultivate it. In fact, one of the
biggest challenges in achieving optimal benefit from this technology is to bring
about the synergistic combination of the skills of the biologist, the biochemist, and
the bioprocess engineer.

The first rDNA product to be approved, human insulin, provides an illustrative
example. By today's standards, the first manufacturing process, developed in the
early 1980s, was quite primitive. Insulin is a protein composed of two polypeptide
polymers connected by disulfide bonds. The first production process was initiated
at Genentech, Inc., and was improved and implemented at Eli Lilly and Co. The
process produced each polypeptide separately. First, the individual polypeptides
were expressed and accumulated inside an engineered bacterium. However, the
polypeptides expressed by themselves were not stable in exposure to the degradative
enzymes found inside the bacterium, so they had to be expressed as a small portion
of a much larger molecule. The large fusion protein then aggregated into a stable
particle inside the cell. To obtain the desired polypeptide, the particle then had to
be isolated and solubilized and the fusion protein cleaved with a hazardous chemical
that produced many side products. In fact, most of the recombinant protein produced either was not the desired polypeptide or was modified and therefore had to
be discarded. The fraction of the total that was the desired polypeptide then had to
be purified and combined with its partner (which had been produced from another
bacterium in a similar manner), and the resulting insulin molecule was purified
again.

Implementing that process was an impressive bioprocess-engineering accom-
plishment. The first process that duplicated all the manufacturing steps was put into
mass production in 1982; it was soon improved to use the single-step concept based
on proinsulin. The most significant advances were achieved by implementing new
ideas, rather than just implementing careful engineering in the traditional sense of
scaleup and cost reduction. The new process depends on advances in the biology of
protein expression and on new developments in the biochemistry of protein modification. Now both polypeptides are expressed simultaneously in the same bacterium.
They are initially connected by another peptide, which guides their assembly. The
connecting peptide is then removed by an enzymatic treatment—a step that turned
out to be the most difficult task in developing the process. This process is quite
similar to the way that insulin is made naturally. Without a strong partnership of

biochemists, molecular biologists, microbiologists, and bioprocess engineers, this new and efficient process would not have been achieved. Both scientists and engineers were required to make this important advance in manufacturing, with 300 employees at Eli Lilly laying the groundwork (in 1986) for mass-producing biotechnology products (Eli Lilly and Company, 1986).

A similar story can be told about the development of human growth hormone. The first rDNA human growth hormone was made with a process in which the protein accumulates in the bacterium. Large quantities of bioactive growth hormone are obtained in this way, but an artifact of intracellular expression produces a molecule with one extra amino acid (192 instead of 191). Although the enlarged product proved to be safe and effective, a growth hormone with the authentic human amino acid sequence was desired. Both Genentech and Eli Lilly were able to make it. It is not publicly known how Eli Lilly accomplished this for its commercial process, but its publications suggest that it produced a fusion protein in the bacterium and then specifically removed the amino acid extension to produce the authentic growth hormone sequence. Genentech solved the problem by engineering the bacterium to transport the growth hormone from the bacterium's central compartment into another portion. As a result of the process of translocation, the correct molecule is formed. Again, by more closely mimicking the natural production mechanism, a superior production process and an improved product resulted.

When the opportunity arose to develop tissue plasminogen activator (t-PA) as an important treatment for heart-attack patients, bacterial expression was again evaluated. This time, the molecule was too large and complex. Even with rapidly developing bacterial technology, a combination of altered expression and protein modification could not reliably produce authentic bioactive t-PA. But the ability to express foreign proteins in mammalian cells had just been developed, and human t-PA was expressed in an active form by cultured, rDNA-modified Chinese hamster ovary cells. The experience with erythropoietin (EPO) was similar, but the technology for large-scale culture of mammalian cells was in its infancy. In fact, the current situation is surprisingly similar to that in the early days of the antibiotic industry 50 years before. Just as the early antibiotic-producing cultures were grown in banks of milk bottles, so the new rDNA mammalian cells were grown in banks of bottles called roller bottles, although robotics (not available in 1943) are used in the modern EPO process.

The amount of EPO needed is small, because EPO is effective in minute doses. Hence, EPO production was scaled up simply by automating and expanding the roller-bottle facility. Amgen and its Japanese partners used an approach relying predominantly on traditional engineering. Genentech, however, in developing t-PA, was developing a product that required a much larger dose per treatment. Not only did the cost per unit of protein need to be proportionately lower, but the production requirements were also much larger. Thus, an alternative approach was required. Just as the early antibiotic industry used larger stirred vessels to make more product at a lower cost, so Genentech developed a process that allowed the production of t-PA in large stirred vessels. However, it was not a matter of merely placing the cells in a larger vessel. The first cultured cells grew optimally only when they were attached to a solid surface. Engineering solutions were available, but the best solution was to select new cells that not only made more t-PA, but also grew well when suspended in a liquid culture medium. Many other improvements were needed to produce a workable mammalian cell-culture process for t-PA, and engineering was a critical partner in the scaleup activity. Most of the additional improvements also resulted from the combined efforts of biologists, biochemists, and bioprocess engineers.

Those examples focus on the part of the production process that first makes the desired protein. But that is only the beginning. The proteins must be purified, not only from the hundreds of other proteins produced by the rDNA organism, but also from variants of the desired protein that differ in subtle but important ways. It takes a major effort merely to be able to detect and measure the two classes of contaminants. One testimony to the success of the purification and analytical efforts is that the contaminating proteins from the producing organism are often measured in the parts-per-million range—a degree of protein purity that was unheard of before the advent of the rDNA-protein pharmaceutical industry made it possible to manufacture proteins in quantities a million times greater than the amounts found in humans. These processes were developed and implemented with a multidisciplinary team approach that used bioprocess engineering in all phases of product development, manufacturing scaleup, and plant operations.

though diverse in biochemical characteristics, must be uniform in its sensitivity to undesired chemical, physical, and enzymatic modification. Furthermore, regulatory and safety demands have required the purification of these protein pharmaceuticals to an unprecedented degree.

The early products selected for development either were quite potent or entered established markets. Examples of marketed rDNA-protein pharmaceuticals are listed in Table 2.1. With the possible exception of insulin (which entered a more mature market and for which some process technology was already established), these proteins have a relatively high unit value and have been needed in small quantities. The first challenge for bioprocess engineering was thus to establish the foundation technologies required to produce protein pharmaceuticals of acceptable quality, albeit at a high cost and on a small scale.

As the industry continues to develop, however, many of the newer product candidates must be manufactured at much lower cost and higher capacity. They include various therapeutic monoclonal antibodies and IGF-1 and range to such items as human serum albumin and human hemoglobin. The latter two products will probably require selling prices of less than \$10/g and manufacturing capacities greater than 10,000 kg/yr. Thus, protein pharmaceuticals have ranges of more than a factor of 10^5 in unit value and projected production volume. For the lower-unit-value products, effective bioprocess engineering might well spell the difference between success and failure. Expansion of the present technology base by well-coordinated, multidisciplinary bioprocess development will be essential for the required reduction in manufacturing cost and increase in capacity.

Table 2-1 Unit Values and Relative Production Quantities
for Selected Approved Biopharmaceuticals

Product	Year Approved	Approximate Selling Price, $/g	Amount of Product for $200 Million in Sales, kg
Human insulin	1982	375	530
Growth hormone	1985	35,000	5.7
Tissue plasminogen activator	1987	23,000	8.7
Erythropoietin	1989	840,000	0.24

Even for products with high unit values, effective bioprocess engineering will be an important competitive factor. Often several companies are in competition to develop the same drug. The company with the first approval or the more attractive product usually has a much stronger competitive position. A distinct advantage will be gained if rapid process development can allow earlier entry into clinical trials. Additional advantage will be gained if the production process results in a product whose biochemical characteristics are more acceptable to the regulatory authorities and to the customer.

In summary, the challenge for U.S. biopharmaceutical companies is to develop quickly the technological tools and the processes that produce superior products at acceptable costs and in the required amounts.

2.2.2 Bioprocess Engineering Requires Many Disciplines

Bioprocess development for biopharmaceuticals involves all aspects of generating a safe, effective, and stable product. It begins with the biological system, continues with product isolation and purification, and finishes when the product is placed in a stable, efficacious, and convenient form. The product is initially derived either directly or indirectly from a living organism. Thus, process development starts with the development of the biological system. It is usually a living organism that expresses the desired protein; but it might be an enzyme for protein modification or an antibody for immunoaffinity purification. rDNA and hybridoma technology allow the biological system to be optimized for maximal formation of the product, for facilitation of downstream processing, for high product quality, and for improved interaction with the production equipment. In this phase of bioprocess engineering, many disciplines must be applied, including molecular biology, genetics, biochemistry, analytical chemistry, and bioprocess engineering. Thus, engineers become full partners with experts trained in the bioscience disciplines in developing and scaling up manufacturing technology for biopharmaceuticals.

As the task of process development proceeds through isolation, purification, and formulation, multidisciplinary approaches continue to be advantageous. Bioprocess engineers combine their skills with those of biochemists and analytical biochemists to develop the optimal process.

The multidisciplinary task of bioprocess engineering is usually applied to developing processes for the production of biopharmaceutical products, but it can also be applied to the discovery of new products. For example, automated screening methods that use purified cell receptors can be developed. More efficient methods for producing and isolating families of precisely modified proteins can be devised, and the technology required to produce protein pharmaceuticals efficiently with mammalian cells can be applied to develop cell-based assays to screen for product activity or toxicity.

2.2.3 Opportunities

The growing biopharmaceutical industry is facing many challenges. Some are technical challenges that can be met by more effective bioprocess engineering or that impede the progress of bioprocess-engineering development. Others are nontechnical, but are strong impediments to the ability of bioprocess engineering to contribute to the competitive position of the U.S. biopharmaceutical industry.

The 1991 Office of Technology Assessment report *Biotechnology in a Global Economy* describes three kinds of research: basic, generic applied, and applied research. From the perspective of the biopharmaceutical industry, the first appears to be well addressed by existing university and government laboratory programs. The last, applied research, is in most cases effectively and appropriately addressed by the private sector during the development of individual products. It is the middle category, generic applied research, that will be most critical for the optimal exploitation of bioprocess engineering in the U.S. biopharmaceutical industry. It is also called "bridge" research, because it bridges the gap between scientific knowledge and practical application. It is often too applied for the scientific disciplines to address, but too risky or requiring too long a period for results for companies to pursue. Table 2.2 lists key technical challenges that the biopharmaceutical industry faces today; most are in the category of generic applied research.

Generic applied research has proved to be most valuable for rDNA-pharmaceutical development. It includes biological and biochemical research that a company cannot afford to do but that fits into the mission of the National Institutes of Health (NIH) in the context of a health agency.

Table 2.2   Key Technical Challenges in Biopharmaceuticals

1. To develop methods for rapid characterization of biochemical properties, efficacy, and immunogenicity of protein pharmaceuticals (methods should be developed to facilitate rapid development of new products and improvement of processes for existing products).
2. To improve process control and process productivity from genetically altered mammalian cell products.
3. To develop high-resolution protein purification technologies that are relatively inexpensive, are easily scaled up, and have minimal waste-disposal requirements.
4. To expand the range of biopharmaceuticals that can be produced with prokaryotic cells or nonmammalian expression systems, which allow the use of less expensive media and have lower capital requirements than current technologies.
5. To expand technology for stable liquid formulations and nonparenteral administration (or sustained release) of protein pharmaceuticals.
6. To increase knowledge of chemical and biochemical reactions that modify proteins during production and storage.

Indeed, in addition to the National Science Foundation (NSF) and other federal agencies that address bioprocess-engineering and biomanufacturing-technology research, NIH has made important contributions to health-product-related, generic applied research in the past, and it is an important component for the future, as indicated in the Federal Coordinating Council for Science, Engineering, and Technology (FCCSET) report (1992).

Table 2.3 lists some nontechnical challenges related to the ability of bioprocess engineering to contribute to the competitiveness of the U.S. biopharmaceutical industry.

Table 2.3   Key Nontechnical Challenges in Biopharmaceuticals

1. To reduce time required for approval of new protein pharmaceuticals and improved processes by improving communication between industry and Food and Drug Administration, reducing response times after submissions, and ensuring timely resolution of generic safety, efficacy, and manufacturing issues.
2. To increase funding opportunities for intensive and sustained technology development required for developing generic applied technology (includes establishing large manufacturing facilities required for relatively large-volume, low-value-added pharmaceuticals).
3. To provide improved training to enhance integration of many disciplines required for optimal bioprocess development and provide training programs specifically to prepare engineering students for biopharmaceutical industry.

2.3 THE ENVIRONMENT

The four major categories of biotechnology applications involved in solving environmental problems are biomaintenance, bioremediation, waste minimization, and environmental monitoring. There are selected opportunities for bioprocess engineering in applying microbe- or enzyme-based treatment protocols to large land masses. Expansion of the knowledge base of different kinds of organisms and their contributions to the ecosystem will help to identify specific mechanisms in the complex, naturally occurring detoxifying activity of our environment. Exploration of biodiversity and use of newly recognized organisms will benefit from this activity. Tremendous engineering challenges must be met if we are to implement rate-enhanced processes in synergy with indigenous phenomena to detoxify the soil, water, and air biologically. The toxins are often dilute and dispersed over a large area. An engineering-bioscience interface will be critical in any biology-based approach, given the need for new genetically designed organisms and systems for cleaning up of site-specific toxins at economically feasible bioremediation facilities. Increased training of microbial and biochemical ecologists, biohydrogeologists, and bioprocess engineers will be useful in providing a cadre of persons to solve these problems. Such an activity, in many cases, would be a logical extension of existing programs and skills in agriculture and civil engineering. It is the committee's opinion that the impact of bioprocess engineering on environmental issues is of great significance and warrants an independent study and analysis.

2.4 CONVERSION OF RENEWABLE AND NONRENEWABLE RESOURCES

Most of the applications and potential applications of bioprocessing related to renewable and nonrenewable resources involve large-scale operations and products of relatively low value. The costs of processing have to be low, and the decision to use bioprocessing for such raw materials must be made with care. Precedents for successful (commercial) large-scale bioprocess-engineered processes include the corn wet-milling industry, the fuel-alcohol industry, the mining industry (specifically copper extraction), the wastewater-treatment industry, the acetone-butanol fermentation industry (stopped in the West in 1955, but still practiced in China), and fermentation industry that manufactures amino acids and other organic acids (including citric and lactic acids).

2.4.1 Renewable Resources

The most abundant renewable material is lignocellulose. Wood, agricultural residue (corn stover, straw, etc.), plants grown deliberately for biomass (such as hybrid aspen), and recycled pulp fiber are the main sources of lignocellulose. Its largest industrial use is in making pulps for paper and other fiber products; second is the use of wood directly in construction. Most current pulping processes use mechanical or chemical processing of lignocellulose. A few bioprocesses are used with lignocelluloses including ensiling and the retting of bast fibers. Two new bioproducts for the pulp and paper industry have reached the market in the last 2 years: a fungal inoculum used to depitch wood chips before pulping, and a cellulase-hemi-cellulase enzyme mixture for improving pulp drainage in paper-making. There is considerable room for additional applications of bioprocessing (Kirk and Chang, 1990). Among applications being researched are the use of enzymes for bleaching of chemical pulps, in the processing of recycled fibers, and in enhancing the surface properties of mechanical and chemical pulps; biopulping (the use of lignin-degrading fungi to soften wood chips for pulping); improvements in waste treatment; fermentation of hydrolysates, especially to ethanol; and bioconversion of manufacturing byproducts to value-added uses. Ethanol production is discussed in more detail later.

The development of inexpensive fermentable sugars would promote growth of a fermentation and chemical-conversion industry based on renewable resources. That would open the way for production of specialty chemicals that have values of $1-10/kg and also aid development of advanced technologies applicable to higher-value products. Examples of chemicals are organic acids (lactic and butyric), ketones (acetone and methyl ethyl ketone), alcohols (ethanol, butanol, and isopropanol), 2,3-butanediol (a precursor of 1,3-butadiene), microbial polysaccharides (for drilling oil wells), and amino acids (for animal feeding) (Jain et al., 1989; Bungay, 1992). Examples of higher-value products would be flavors, fragrances, pigments, and pharmaceuticals.

Starch derived from corn and other crops is an abundant renewable material. During the last decade, industrial uses of starch have more than doubled. The U.S. corn wet- and dry-milling industries processed over 1.2 billion bushels of corn (approximately 15% of the annual crop) in the year starting in September 1990 to make a variety of products for food and industrial uses, including high-fructose corn syrup (HFCS), glucose and dextrose, industrial starches, and fermentation feedstocks (van Meir and Baker, 1992), and approximately 850 million gallons of fuel ethanol were made by the milling industries in 1991 as a gasoline additive. HFCS and ethanol production consumed about 9% of the 1991 corn crop (van Meir and Baker, 1992; Buzzanell et al., 1992); these products are made with processes that integrate enzymatic steps, fermentation, and separation technologies.

Ethanol produced from cellulosic biomass has the potential to provide an important percentage of the U.S. need for liquid transportation fuel (Ladisch and Svarczkopf, 1991; Lynd et al., 1991) while reducing dependence on foreign oil and decreasing the production of fossil-fuel-derived greenhouse gases.

The bioprocess engineer can contribute to research, development, and manufacturing. In the case of cellulose conversion, both pretreatment technology and enzyme hydrolysis are the most expensive steps and so have the greatest potential for improvement. The current cost of enzymes limits the amount that can be used. More efficient fermentation of xylose is also

---

Vignette 3

Pulp and Paper Bioprocessing

Manufacture of pulp and paper, with a U.S. market value of $122 billion a year, is the largest U.S. use of wood. Recent developments in biotechnology have facilitated primary manufacturing, and further developments are anticipated. Two new biotechnology-based products have been introduced in the last 2 years. "Liftase," from Biopulp International in France, is a mixture of enzymes (cellulases and hemicellulases) sold as a drainage aid for paper-making. Chemical pulp, produced by wood delignification, is introduced continuously onto paper machines as a slurry; the wet paper mat's first support is a continuous wire screen through which water from the pulp drains. The sheet of wet paper is further dewatered by being pressed on a fabric belt and then on heated metal rollers; this results in production of a continuous sheet of paper at the other end of the paper machine. Initial drainage of the water through the screen determines the paper-machine speed. The drainage-aid enzymes, sold as a solution and introduced into the pulp slurry, digest water-retaining fiber fines and viscous hemicelluloses, speeding water release and allowing higher paper machine speeds.

The second product, "Cartapip," developed in the United States by the Swiss-U.S. company Repligen-Sandoz Research Corp., removes pitch from wood chips before pulping. *Pitch* is a catch-all term for the mixture of lipids, resins, and other extraneous compounds found in wood. After wood chips are converted into pulp for making paper, pitch is often a serious problem, causing paper to stick to the drying rolls and leading to formation of dark spots and tears. In addition, pitch components in wastewater are often toxic. Classical methods for ameliorating the problems caused by pitch include adding talc to the pulp to coat the pitch. The new biotechnology-based product is a dry powder consisting of inoculum of a natural strain of the fungus *Philiostoma piliferum* in an inert carrier. Formulated as a slurry, the inoculum is sprayed onto wood chips as they are piled outside for storage before being pulped. The fungus grows through the piles within 2 weeks, gaining its nourishment by metabolizing the pitch. The fungus does not affect the wood cellulose and so does not decrease pulp yield or quality.

Bioprocess engineers were instrumental in developing the industrial microbial processes for producing both the enzyme mixture and the fungal inoculum.

needed. A particularly attractive possibility is to combine in a single microorganism the ability to produce extracellular cellulose- and hemicellulose-hydrolyzing enzymes with the ability to ferment the sugar mixture. Advanced bioreactor concepts, especially for high-volume products, might allow significant increases in productivity and thus reductions in capital and operating costs. Research subjects in the engineering of bioreactors include continuous operation, more efficient microbial-solids contact, improved operational control of large fermentors, and manipulation of metabolic pathways to increase microbial productivity.

Some of the major wet millers are moving into other types of fermentation products for which on-site availability of sugars from corn could provide a competitive edge (Table 2.4). Fermentation-derived citric acid, amino acids (particularly lysine), and numerous other products are in various stages of development, and growth is anticipated as new biotechnology processes, relevant to production of moderate-value products, come of age (Anonymous, 1991). Those products affect the energy, environmental, and food-processing sectors (Glaser and Dutton, 1992) of the U.S. economy. The total value of the U.S. wet- and dry-milling output exceeds $4 billion a year ($2.8 billion for HFCS and $1 billion for fuel alcohol), and 97% of these products is consumed domestically. The development of these industries is based on technological contributions from bioprocess engineering.

2.4.2 Nonrenewable Resources

Bioprocessing is used much less with nonrenewable resources than with renewable resources. Nevertheless, bioprocessing has found a few commercial applications, including the leaching of copper from ore by bacteria and treatment of wastewater. Some additional possibilities are being studied (Finnerty, 1992; Lee and Scott, 1988; Watson and Scott, 1988).

Research is under way in a number of federal and university laboratories on bioprocessing of coal, microbial removal of sulfur and other heteroatoms, microbial or enzymatic solubilization or liquefaction, bioconversion of byproduct gases (methane and synthesis gas) to liquid fuels, gasification of coal by one- or two-stage microbial processes, and bioremediation of organic pollutants from coal-conversion wastewater and soils. In the petroleum industry, bioprocesses are now used mainly in treatment of processing wastes. Bioprocessing has potential, however, for enhancing the recovery of oil, remediating soil and water, converting petroleum components to higher-value products (including synthons), controlling fouling in processing plants, and providing biosurfactants for a variety of applications.

Vignette 4

**Bioprocess Engineering for High-Volume Products: The Case of Corn
and the Wet-Milling Industry**

High-volume products from corn use bioprocess-engineering principles in applying enzymes, fermentation, and separations for production of industrial sugars and
fuel alcohol.   Fermentation alcohol and enzyme-based production of HFCS predate
commercial ventures based on the new biotechnology and provide case studies for
large-scale adaptation of bioprocesses in an established field where competitive
options are readily accessible.   The enabling discovery of xylose (glucose) isomerase
was based on classical microbiology and enzymology, rather than recombinant DNA,
but the history of its translation into a large-scale manufacturing technology provides useful corollaries for events that might shape the bioprocessing industry in the
next 10 years.

The glucose-isomerizing capability of xylose isomerase, an enzyme that had
been discovered in 1954, was observed at the U.S. Department of Agriculture (USDA)
Peoria laboratory in 1957 (Marshall and Kooi, 1957).   However, arsenate was required as a cofactor, and this enzyme was later found to be a glucose phosphate
isomerase, rather than a glucose isomerase.   Other enzymes that did not require
arsenate for activity were quickly identified by researchers in Japan, and that opened
the possibility of food use.   Clinton Corn Processing Company of Iowa, in 1966,
entered into an agreement with the Japanese government, commercialized a glucose-
isomerizing process, and shipped the first enzymatically produced 42% fructose
syrup by 1967.   A. E. Staley licensed technology from Clinton and also entered the
market.   By 1974, U.S. production grew to 500 million pounds a year as sugar prices
peaked at $0.30/lb, thereby encouraging investment in new plants.

The Japanese technology made it possible to produce xylose isomerase from
Streptomyces rubiginosus, using corn bran as an economical source of xylose to
induce the microbial production of the enzyme.   The resulting enzyme was thermally stable and had high glucose-isomerizing activity at the industrially relevant conditions of 65°C and pH 7.3 (Lloyd and Horwath, 1985).   Selection of the appropriate
microorganism and adaptation of media composition facilitated the enzyme's economical production.

The first processes consisted of adding enzyme to batches of glucose syrup.   The
recognition that glucose isomerase could be immobilized by sorption on DEAE
cellulose or by entrapment in heat-fixed, pelletized cells (the enzyme has a molecular weight of about 170,000) improved production economics by providing stable,
reusable enzyme (Lloyd and Horwath, 1985; Lloyd and Khaleeluddin, 1976).   The
current industrial practice is to fix or entrap the enzyme on a solid material, which,
in turn, is packed into a fixed bed.   The glucose syrup, derived from enzyme hydrolysis of corn starch, is then passed over a fixed bed of immobilized isomerase, which
converts the glucose to fructose.   The current practice owes much to bioprocess
engineering, including technology for enzyme immobilization, bioreaction engineering and kinetics, development of physically and chemically stable supports (onto
which the enzyme is fixed), and optimal design and operation strategies for a protein-based catalyst.

The basic patent coverage for use of xylose isomerase to convert glucose to
fructose was held invalid in 1975.   That provided incentives to develop alternative
processes and process improvements and was followed by an industry-wide, major

construction boom. Next came overcapacity and low profitability in 1977, when sugar prices fell to $0.12/lb and annual U.S. production had grown to over 3 billion pounds. Those events could find analogies in future developments of the biopharmaceutical industry, where process innovations should be anticipated.

Separation technology was also important in the growth of HFCS production. The introduction of very-large-scale liquid chromatography enabled partial separation of glucose from fructose and the production of 55% HFCS (abbreviated HFCS-55). HFCS-55 had sweetening properties nearly equivalent to those of inverted (hydrolyzed) sugar from sucrose and began to appear in soft drinks in 1980. HFCS syrups, based principally on corn grown in North America, now dominate the industrial sugar market, with about 12.5 billion pounds of 42% and 55% HFCS shipped in 1991 in the United States (Buzzanell et al., 1992). Between 1978 and 1990, the industry grew by 350%.

The HFCS industry, with its large scale of operation and infrastructure, was largely responsible for the development of large-scale production of ethanol, another product derived from starch, and particularly from corn. Fuel-ethanol production involves fermentation of glucose to ethanol with the yeast *Saccharomyces cerevisiae* and requires very large fermentors. A cascade system, now operated by most large U.S. plants, evolved in which there is continuous flow through a series of tanks, approximating a continuous fermentation (Hacking, 1986). After fermentation, the solids are separated from the fermentation broth, dried, and sold as animal feed. Recovery of ethanol entails distillation to approximately 92% alcohol followed by azeotropic or extractive distillation. Discovery of corn as a suitable drying agent for alcohol in 1979 (Ladisch and Dyck, 1979) and its later adaptation on a commercial scale by a major alcohol producer in the mid-1980s provided an environmentally acceptable and energy-efficient method of using corn in a manner compatible with a fixed-bed sorption process for large-scale alcohol-drying (Lee et al., 1991).

Fundamental research in ethanol production was encouraged through federal funding of research through the Department of Energy (DOE), USDA, and NSF from about 1975 to 1985. That period saw examination of the engineering, enzyme technology, biological science, and economics that would facilitate economic conversion of cellulose to various oxygenated chemicals (including ethanol) via fermentable sugars. Direct fermentation of cellulose to ethanol was proved to be technically feasible (Wang et al., 1983). Complete conversion of cellulose to sugars via enzyme hydrolysis of pretreated biomass was demonstrated (Ladisch et al., 1978). Recombinant methods and applied microbiology resulted in large gains in microbial productivity and cellulose saccharification activity of cellulolytic enzymes (Mandels et al., 1981; Montenecourt et al., 1981). Organosolv pretreatments further enhanced hydrolysis prospects (Holtzapple and Humphrey, 1984; Avgerinos and Wang, 1983). The complex mechanisms of lignin degradation were elucidated (Tien and Kirk, 1984). The potential of xylose isomerase in fermenting pentoses (a major constituent of cellulosic materials) to ethanol was shown in a fermentation process (Gong et al., 1981). Elucidation of the complex metabolism associated with ethanol fermentation resulted in methods for improving ethanol-fermentor design (Maiorella et al., 1984) and bacterial fermentation of pentoses to ethanol (Ingram, 1992).

Only a small part of that research, however, found its way into industrial practice, partly because the precipitous decrease in oil prices from $40/barrel to $20/barrel in 1985-1986 removed many of the visible, short-term economic and strategic incentives for moving rapidly to a renewable-resource-based industry. Nonetheless, the period fostered training of bioprocess engineers, who received a cross-disciplinary education as part of research projects related to renewable resources. Many of them joined the growing biopharmaceutical industry in bioprocess-engineering capacities.

2.4.3 Coupled Synthesis Gas

One strategy for using biomass in fuel and chemical production is to reform it thermally to synthesis gas and then to use the synthesis gas in existing petrochemical facilities. The development of appropriate technology for such a process will require elements of bioprocessing and biology. The development of biomass sources that are most appropriate for reforming, converting the biomass to the physical form most appropriate for reforming, and reuse or disposal of the byproducts of the process will require an infusion of bioprocess-engineering concepts.

2.4.4 Enhanced Oil Recovery

Microbial polysaccharides (xanthan and guar gums and related polysaccharides) are an important component of enhanced oil recovery (EOR). There is an opportunity to develop new additives for down-hole viscosity

Table 2.4   Biochemical Processing Related to the Corn-Refining Industry

Product	Company	Capacity, tons/yr
Vitamin C	Takeda Chemical Products, United States	5,500
Enzymes	Genencor International (Cedar Rapids, Iowa)	($60 million facility)
Citric acid	Cargill (Eddyville, Iowa)	27,500
Citric acid	ADM (Decatur, Ill.)	Not known
Lysine	Heartland Lysine (Eddyville, Iowa)	20,000
Lysine	Biokyowa (headquarters at Chesterfield, Mo.)	13,000
Lysine	ADM (Decatur, Ill.)	62,500
Tryptophan and threonine (planned)	ADM (Decatur, Ill.)	Not known
Lactic acid, feed-grade penicillin and bacitracin, and biotin (planned)	ADM (Decatur, Ill.)	10,000-20,000 (planning stages)

Source: Anonymous, 1991.

modification through genetic engineering, to develop on-site production techniques for remote environments, and to develop techniques for other aspects of EOR (surfactants for heavy-crude slurry transport and site remediation). These problems can be approached through bioprocess engineering.

2.4.5 Opportunities

The bioprocessing of renewable and nonrenewable resources requires the application of multiple disciplines in a cross-disciplinary approach. As in the case of biopharmaceuticals, products are derived either directly or indirectly from living organisms. Unlike those in the biopharmaceutical sector, however, bioprocess engineers dealing with renewable and nonrenewable resources are involved with high processing volumes and often design and operate large installations. The processes themselves must still be reliable, robust, and cost-effective. The goal is to generate products from renewable resources or to produce microbial and other biocatalysts for use in processing renewable or nonrenewable resources to obtain value-added products. Economies of scale and scalable equipment are important, if not critical, process characteristics.

This sector of the biotechnology industry also produces specialty products (enzymes, amino acids, organic acids, and animal-health products) for which the processing technology and regulatory issues are similar to those associated with high-volume biopharmaceutical products, but the ratio of market price to processing cost can be much lower. Production efficiency, as well as product quality, is paramount. A key function of bioprocess engineering is the design, operation, and improvement of manufacturing technology to yield bioproducts at high volume and acceptable costs. Success will require bioprocess engineers to become partners with molecular biologists and geneticists (both plant and microbial), biochemists, carbohydrate chemists, wood chemists, food scientists, analytical chemists, and microbial physiologists.

Substantial generic applied research is needed to help foster the development of the U.S. bioprocessing industry for manufacture of specialty products. As listed in Table 2.5, the technical challenges will require cooperative efforts among university researchers, industry, and national government laboratories and research institutions of USDA, DOE, the Environmental Protection Agency, the National Aeronautics and Space Administration (NASA), the National Institute of Standards and Technology, and the Department of Defense (DOD). The challenges listed in Table 2.5 present opportunities related to renewable resources to carry out generic applied research in the universities and government laboratories, because these technologies will have wide impacts on the use of land and natural resources.

Table 2.5   Key Technical Challenges in Bioprocessing of Renewable
Resources

1. To develop inexpensive cellulose pretreatment and saccharification processes effective
   with lignocellulosic materials on large scale with environmentally compatible methods.
2. To develop fermentations capable of converting pentoses to value-added products at
   yields, rates, and extents similar to those obtained for glucose with yeast and to increase
   product concentrations achievable in both hexose and pentose fermentations.
3. To develop more efficient separations for recovering fermentation products, sugars, and
   other dissolved materials from water, i.e., lower cost of separating water from product in
   fermentation broth.
4. To develop processes for large-scale inoculation, control, and propagation of
   microorganisms in surface culture (e.g., treatment of wood chips and bioremediation of
   soils) and solid substrate fermentation.
5. To increase knowledge of combinations of chemical, biochemical, and microbial
   transformations that result in value-added nonfood products from starch and cellulose.
6. To improve fractionation methods for separating oil, starch, and fiber components during
   corn milling to obtain higher coproduct values with lower capital investment.

The universities also have the important task of training bioprocess
engineers who will be both competent in engineering fundamentals and
conversant in the life sciences.   Success will require a cross-disciplinary
approach.   All three groups—industry, government laboratories, and universities—will need to seek partnerships to build the first pilot units for new
bioprocessing technologies.   In many instances, the scale required for carrying out worthy fundamental research will be quite large, and realistic feed
stocks will be available only at the plant site.   As a consequence, special
efforts will be needed to transfer technology from the laboratory to the pilot
scale and ultimately to full-scale production.

Industry involvement and the special facilities and capabilities of government laboratories could rapidly achieve the first adaptation of new bioprocesses on a commercial scale.   A historical precedent is found in the
development of penicillin (see Vignette 1).   Over the last 10 years, the NSF
Engineering Research Centers program and more recently the NASA Scientific Centers for research and training have provided working relationships
of this type.   The committee recommends that these models be examined
and applied, in a suitably modified form, to the processes for obtaining
value-added products from renewable resources.

Incentives for adapting bioprocess technology might be provided by environmental concerns, government tax policies, the possibility of improving
product quality, and economic factors.   The first investments in new bioprocess technology might appear to yield small unit returns (relative to a high-
value-added product), but the overall volume of product will likely be immense.   The effects on the U.S. economy would be significant, particularly

if biobased products with improved end-use properties begin to supplant products derived from petrochemical sources and methods of producing and processing food are changed through bioprocessing. The committee recommends that a separate study on biobased materials be considered to explore the potential impact of adaptation of bioprocessing technology on the input-output grid of the U.S. economy.

2.5 SPACE

Bioprocessing has two major categories of applications related to space: manufacture of biomaterials and bioregenerative life-support systems. Both require bioprocess engineers who are conversant in fundamentals of biochemical engineering, including physical and chemical properties of biomolecules, bioreactor design principles, separations, fluid flow, heat transfer, and basic biochemistry and biology. Manufacture of biomaterials in space is of little short-term economic interest. The space environment, however, does provide a laboratory for developmental biology and physiology, as discussed in Section 6.1.13.

An important feature of bioprocessing in space that distinguishes it from other types of bioprocessing is its application in a systems environment. Reliability and safety are paramount, because of the hostile and remote location of a space station or colony and transportation costs of $8,000-10,000/kg, which limit resupply options. Integration into other subsystems not related to bioprocessing must be carried out. Consequently, bioprocess engineers with a systems background are needed. The scale of staffing needs might be understood if it is considered that several thousand engineers are at work on the space station. Very few have a bioprocess-engineering background, but the space station or a Moon or Mars colony will require bioprocess engineers as strategic members of the various aeronautical-, mechanical-, thermal-, and structural-design teams to foster communication and enable evaluation of bioprocess options for bioregenerative life support.

The national space-policy goals are to return to the Moon and establish a permanent human presence. The space station will serve as an international laboratory for study of plants and bioregenerative systems, as well as the ultimate biological system, humans, before long-term trips away from Earth are carried out. The residence time of humans in space will be long, relative to previous experience (e.g., 84 days in the Skylab in 1974). To achieve a Moon base by 2010 and a Mars base by 2020, a manned space station is to be operational by the year 2000. The Moon is intended as a possible staging area for exploration and colonization of Mars. Human travels outside the Earth's biosphere are envisioned to be about 180 days for the Moon base and about 3 years for Mars. Systems are needed that are extremely reliable and function with minimal resupply (because of the risk and ex-

pense associated with current space transportation systems). Regenerative
life support is a critical need and will likely be based on a combination of
physicochemical and biological subsystems. Modeling of the various components of bioregenerative life support is needed to be able to predict responses of an ecosystem to changes in operating conditions, throughputs, or
environmental conditions.

Bioprocess engineering is a critical element in developing subsystems
and integrating biological and mechanical subsystems for bioregenerative
life support, for which the criterion of success is reliability, rather than
economics. There is no technology base that NASA can rely on to develop
bioregenerative life-support systems. Private-sector initiatives are unlikely
in the near future, because the market (i.e., NASA) for such a system is so
limited. Much of the initial research is being done by plant physiologists
and members of allied professions. It is estimated that 50-100 bioprocess
engineers will be needed by NASA and NASA contractors by the year 2000
to serve as an interface for life-science developments to be incorporated
into component, subsystem, and system designs.

2.6 BIOTECHNOLOGY-RESEARCH INITIATIVE GIVEN BY FEDERAL COORDINATING COUNCIL FOR SCIENCE, ENGINEERING, AND TECHNOLOGY (FCCSET)

A recent report of the FCCSET Committee on the Life Sciences and
Health describes the objectives, funding levels, and agencies of the U.S.
government's biotechnology-research framework (FCCSET, 1992). The goal
of the initiative is to sustain and extend U.S. leadership in biotechnology
research for the twenty-first century. The strategic objectives (p. 2) are to
"extend the scientific and technical foundations for the future development
of biotechnology; ensure the development of the human resource foundations for the future development of biotechnology; accelerate the transfer of
biotechnology research discoveries to commercial applications; and realize
the benefits of biotechnology to the health and well-being of the population
and the protection and restoration of the environment." The report suggests
directions for future efforts that will draw on advances in modern biotechnology based on fundamental research supported by U.S. federal agencies,
most notably NIH.

The greatest impact of biotechnology, thus far, is in human health; future
developments are viewed as likely to lead to substantial reductions in medical costs through advances in prevention, early diagnosis, and treatment.
The report also projects that biotechnology would soon have other impacts:
"The next decade should see unprecedented applications of biotechnology
to agriculture and aquaculture, to the restoration and protection of the environment, to the production of chemicals and fuels, and to many other ar-

eas." The critical role for the federal government, academe, and industry is to ensure "an uninterrupted flow of new ideas and new techniques to U.S. industry," to address the challenge to the U.S. position of international leadership.

The FCCSET report gives a comprehensive overview of the biotechnology research programs of 12 federal agencies and how they are related to each other. A partial list of priorities in the report includes maintaining strong support for broadly applicable foundation research to sustain the momentum of progress in all fields of biotechnology and for health-related fields to capitalize on the opportunity for applications to human health, substantially increasing and closely coordinating the federal investment in biotechnology research related to manufacturing and bioprocessing, strengthening and expanding interdisciplinary research, and increasing training programs at all levels to provide essential human resources for biotechnology development. The present committee further addressed those issues as related to bioprocess engineering and the resources required for developing capabilities in manufacture of biotechnology products. The importance of bioprocessing in realizing the commercial benefits of biotechnology are clearly recognized in the FCCSET report, and the priorities identified by FCCSET overlap with the findings in the present report. The bioprocessing research programs of NSF, USDA, DOD, DOE, NASA, the Department of Health and Human Services, the Department of Commerce, and the Department of the Interior are currently projected at $123.8 million for FY 1993. Those programs must be sustained to continue support of the bioprocessing-research infrastructure.

2.7 SUMMARY

The challenge of bioprocess engineering lies in identifying the needs of the industry and promoting technology transfer and training of engineers who will fit the wide range of markets, activities, and products that are being encompassed by biotechnology and bioprocess-engineering developments. The synthesis and innovation to develop the enabling technologies for industry to exploit the potential of modern biology and chemistry fully is a significant part of this challenge. The vignettes in this chapter illustrate how bioprocessing technology facilitates amplification of existing microbial products, application of new manufacturing techniques for existing products, and the production of proteins derived from mammalian systems that would otherwise not be available in sufficient quantities for practical use as therapeutic agents.

The optimistic projection that the biotechnology industry will grow by a factor of at least 10 over the next 10 years is justified by the pervasive influence of biological processes on everyday aspects of life and by the many possible applications of the discoveries emanating from the life sciences. Bioprocesses require manufacturing skills that span products whose prices range over a factor of 10^7. The definitions of the kinds of expertise that are required for bioprocessing engineering are similarly broad.

The basic principles of biotechnology manufacturing processes are based on understanding of the microorganisms or biocatalysts involved: ensuring consistent product quality and product safety, regardless of scale; addressing environmental consequences of the manufacture and use of the products; and continually improving and developing processes in response to the competitive pressures generated by consumer markets. The engineering operations that cut across those requirements include the development of bioreactor technology, the processing of impure dilute product streams into purified and concentrated form, and the ability to apply engineering economics in decision-making processes during the design, development, operation, and improvement of bioprocess facilities. That requires engineers who can solve engineering problems on the basis of an understanding of the biochemistry and genetics of living organisms. A sustained policy is needed to foster development of a fundamental knowledge base for the manufacture of a broad spectrum of bioproducts and the training of bioprocess engineers who will grow with the growing industry.

2.8 REFERENCES

Abelson, P. H. 1992. Biotechnology in a global economy [editorial]. Science 255(5043):381.

Aiba, S., A. E. Humphrey, and N. F. Millis. 1973. Biochemical Engineering, 2nd Ed. New York: Academic. 434 pp.

Anonymous. 1991. Corn refiners push into biochemical processing. Pp. 15-17 in Corn Annual. Washington, D.C.: Corn Refiners Association.

Antrim, R. L., W. Colilla, and B. J. Schnyder, 1979. Glucose isomerase production of high fructose syrups. Pp. 97-156 in Applied Biochemistry and Bioengineering, Vol. 2, L. B. Wingard, Jr., E. Katchalski-Katzir, and L. Goldstein, eds. New York: Academic.

Avgerinos, G. C., and D.I.C. Wang. 1983. Selective solvent delignification for fermentation enhancement. Biotechnol. Bioeng. 25:67-83.

Bungay, H. 1992. Product opportunities for biomass refining. Enzyme Microb. Technol. 14:501-507.

Burrill, G. S., and K. B. Lee, Jr. 1991. Biotech '92: Promise to Reality. An Industry Annual Report. San Francisco: Ernst and Young.

Buzzanell, P., F. Gray, R. Lord, and W. Moore. 1992. Sugar and Sweetener Situation and Outlook Report. U.S. Department of Agriculture Economic Research Service SSRV17N1. Washington, D.C.: U.S. Department of Agriculture.

Eli Lilly and Company. 1986. Leadership in biotechnology—Lilly biosynthetic manufacturing team improves humalin production process. Eli Lilly and Company Second Quarter Report. Indianapolis: Eli Lilly and Company.

FCCSET (Federal Coordinating Council for Science, Engineering, and Technology). 1992. Biotechnology for the 21st Century. A Report by the FCCSET Committee on Life Sciences and Health, Office of Science and Technology Policy, Executive Office of the President, Washington, D.C.

Finnerty, W. R. 1992. Fossil Energy Biotechnology: A Research Needs Assessment Final Report. DOE Contract 01-91ER30156. Washington, D.C.: U.S. Department of Energy.

Glaser, V., and G. Dutton. 1992. Food processors seek to adapt bioproducts for large-scale manufacturing. GEN 12(2):6-8.

Gong, C.- S., L.- F. Chen, M. C. Flickinger, L. C. Chiang, and G. T. Tsao. 1981. Production of ethanol from D-xylose using D-xylose isomerase and yeasts. Appl. Environ. Microbiol. 41(2):430-436.

Hacking, A. J. 1986. Pp. 214-221 in Economic Aspects of Biotechnology. New York: Cambridge University Press.

Holtzapple, M. T., and A. E. Humphrey. 1984. Effect of organosolv pretreatment on the enzymatic hydrolysis of poplar. Biotechnol. Bioeng. 26:670-676.

Ingram, L. 1992. Genetic engineering of novel bacteria for the conversion of plant polysaccharides into ethanol. Pp. 507-509 in Harnessing Biotechnology for the 21st Century, M. Ladisch and A. Bose, eds. Washington, D.C.: American Chemical Society.

Jain, M. K., R. Datta, and J. G. Zeikus. 1989. High value organic acids fermentation— Emerging products and processes. Pp. 367-389 in Bioprocess Engineering. T.K. Ghose, ed. London: Horwood.

Kirk, T. K., and H.-M. Chang, eds. 1990. Biotechnology in Pulp and Paper Manufacture: Applications and Fundamental Investigations. Boston: Butterworth-Heinemann. 696 pp.

Ladisch, M. R., and K. Dyck. 1979. Dehydration of ethanol: New approach gives positive energy balance. Science 205:898-900.

Ladisch, M. R., and J. A. Svarczkopf. 1991. Ethanol production and the cost of fermentable sugars from biomass. Bioresour. Technol. 36:83-95.

Ladisch, M. R., C. M. Ladisch, and G. T. Tsao. 1978. Cellulose to sugars: New path gives quantitative yield. Science 201:743-745.

Lee, D. E., and C. D. Scott. 1988. Impact of Biotechnology on Coal Processing. Oak Ridge National Laboratory Report ORNL-6459. Oak Ridge, Tenn.: Oak Ridge National Laboratory.

Lee, J. Y., P. J. Westgate, and M. R. Ladisch. 1991. Water and ethanol sorption phenomena on starch. AIChE J. 37:1187-1195.

Lloyd, N. E., and R. O. Horwath. 1985. Biotechnology and the development of enzymes for the HFCS industry. Pp. 116-134 in Bio Expo '85, O. Zaborsky, chairman. Stamford, Conn.: Cahners Exposition Group.

Lloyd, N. E., and K. Khaleeluddin. 1976. A kinetic comparison of *Streptomyces* glucose isomerase in free solution and adsorbed on DEAE cellulose. Cereal Chem. 53(2):270-282.

Lynd, L. R., J. H. Cushman, R. J. Nichols, and C. E. Wyman. 1991. Fuel ethanol from cellulosic biomass. Science 251:1318-1323.

Maiorella, B. L., H. W. Blanch, and C. R. Wilke. 1984. Economic evaluation of ethanol fermentation processes. Biotechnol. Bioeng. 26:1003-1025.

Mandels, M., J. E. Medeiros, R. E. Andreotti, and F. H. Bissett. 1981. Enzymatic hydrolysis of cellulose: Evaluation of cellulase culture filtrates under use conditions. Biotechnol. Bioeng. 23:2009-2026.

Marshall, R. O., and E. R. Kooi. 1957. Enzymatic conversion of D-glucose to D-fructose. Science 125:648-649.

Montenecourt, B. S., S. D. Nhlapo, H. Trimio-Vasquez, S. Cuskey, D.H.J. Schamhart, and D. E. Eveleigh. 1981. Regulatory controls in relation to overproduction of fungal cellulases. Basic Life Sci. 18(Trends Biol. Ferment. Fuels Chem.):33-53.

OTA (Office of Technology Assessment). 1991. Biotechnology in a Global Economy, B. Brown, ed. Office of Technology Assessment, U.S. Congress, Report No. OTA-BA-494. Washington, D.C.: U.S. Government Printing Office.

Shuler, M. L., and F. Kargi. 1991. Bioprocess Engineering: Basic Concepts. Englewood Cliffs, N.J.: Prentice Hall. 448 pp.

Tien, M., T. K. Kirk. 1984. Lignin-degrading enzyme from *Phanerochaete chrysosporium*: Purification, characterization, and catalytic properties of a unique H_2O_2-requiring oxygenase, Proc. Natl. Acad. Sci. USA 81:2280-2284.

van Meir, L., and A. Baker. 1992. Revisions in estimates of food and industrial use for feed grains. Pp. 20-25 in Feed Situation and Outlook Yearbook. U.S. Department of Agriculture Economic Research Service FdS-32. Washington, D.C.: U.S. Department of Agriculture.

Wang, D.I.C., G. C. Avgerinos, I. Biocic, S.-D. Wang, and H.-Y. Fang. 1983. Ethanol from cellulosic biomass—Direct conversion using a mixed culture of *Clostridium thermosaccharolyticum*. Philos. Trans. R. Soc. London B 300:323-333.

Watson, J. S., and C. D. Scott. 1988. The Impact of Bioprocessing on Enhanced Oil Recovery. Oak Ridge National Laboratory Report ORNL/TM-10676. Oak Ridge, Tenn.: Oak Ridge National Laboratory.

3

Benchmarking: Status of U.S. Bioprocessing and Biotechnology

Bioprocess engineering enables translation of biotechnology into products that benefit society. Biotechnology-derived products are an important source of revenue and commercial growth throughout the world and hence are related to issues of international competitiveness. In this context, the Committee on Bioprocess Engineering generated a comparison of the state of U.S. biotechnology with that of Japan and Europe based on the recently published Japanese Technology Evaluation Center (JTEC) panel report on Bioprocess Engineering in Japan sponsored by the National Science Foundation (JTEC, 1992).

3.1 BIOPROCESS ENGINEERING IN JAPAN

Bioprocess engineering in Japan has changed substantially during the last 10 years as a result of Japan's entry into applications of bioprocessing to higher-value products obtained through recombinant-DNA and cell-culture technology. This change was achieved differently in Japan and the United States. In the United States, the new biotechnology is pursued by both large and small (startup) biotechnology companies, but the small companies are virtually nonexistent in Japan, research and development being conducted primarily in large companies. Japanese beverage, food, and pharmaceutical companies are diversifying into high-value products—a trend just now being initiated in the United States. Chemical, polymer, steel, and electronic companies in Japan have also initiated moves into the pharmaceutical sector. Industry-government relationships in Japan are different from those in the United States. In many instances, the government has provided both directions for research and development and financial support to a group of companies to pursue projects that it believes are critical for the commercial development of biotechnology, e.g., in cell culture, bioreactor design, membrane processing, and protein engineering.

The types of products derived from the new biotechnology in Japan are the same as those already developed in the United States (JTEC, 1992). They include interferon, growth hormone, tissue plasminogen activator, erythropoietin, and interleukins. The bioprocess know-how is often similar to that in the United States, inasmuch as the process technology is in many cases licensed from U.S. companies.

Priority is given to process development for the fine tuning of fermentation processes through strain improvement and medium formulation. In contrast, fundamental engineering research appears to be secondary in the university, government, and industry sectors, with a major focus on applied research. The acceptance of and rewards for performing applied research in all sectors are apparent; both government and industry provide financial support to universities to perform applied research. Another element of Japan's applied research is technology transfer, which is accomplished by placing industrial personnel in the university and government laboratories both in Japan and abroad.

Japan appears to have a leading position in automation for both upstream and downstream processing, with a focus on eliminating or reducing the human-process interface, and in hardware development to effect automation that will ensure product quality. Japanese upstream bioprocess engineering is also heavily involved in strain selection, medium development, and environmental control—particularly strain selection, which is believed to be a major spinoff from the amino acid and antibiotic industries. Control of the production environment through medium development is given far more attention than the implementation of completely new engineering concepts. Personnel performing research in the upstream sector are mostly from the departments of agricultural chemistry, fermentation technology, applied biochemistry, and biological chemistry, rather than chemical engineering.

Research with recombinant organisms for the production of biologics from prokaryotes, such as E. coli, does not appear to be very different from that in the United States, although there is a heavy emphasis on the refolding of recombinant proteins from E. coli. Substantial activity in Japan is directed to exploring animal cells for the production of therapeutic proteins. However, innovative bioprocess research on the upstream side is not noted. For example, the animal-cell bioreactors are similar to their U.S. counterparts. In contrast, there is observable and intense activity in animal-cell culture medium development, possibly as part of a national effort to reduce the raw-material cost for animal-cell cultivation.

Japan also has a notable emphasis on biosensor development. Again, research and development focus is on known concepts (e.g., related to enzymes, microorganisms, and antibodies) to be incorporated into biosensor development. Some interesting trends in biosensor research in Japan are reflected in the miniaturization of biosensors with microelectronic technology and the development of disposable and discrete sensors for consumer-oriented biomedical applications.

In downstream processing, as in upstream processing, the development of new concepts or principles is secondary. Most of Japan's bioprocess-engineering activities are devoted to development and refinement of existing technologies, many of which are based on technology developments in the United States and Europe.

The U.S. lead in process validation and Food and Drug Administration (FDA) compliance reflects a serious commitment of time and funds. Japan recognizes the importance of this area for the commercial development of biotechnology, although much of its present knowledge in this field is obtained through technology licenses from the United States.

3.1.1   Education and Training

Japan has a strong educational and training program in applied biology and chemistry. Its strength in traditional fermentation technology results from a long-standing emphasis on applied biology, including agricultural biochemistry, industrial chemistry, and fermentation technology. Continued emphasis on strengthening applied biology seems to support the effective transfer of technology from basic molecular biology to commercially important biotechnology, as well as the industry's success in many branches of bioprocess technology.

3.1.2   University-Industry Cooperation

Japanese industry has supported university research in numerous cases, largely through long-term grants-in-aid. With that kind of support, universities have been able to conduct much long-term exploratory research and to support graduate students and postdoctoral trainees on a sustained basis.

3.1.3   Scientific and Technological Information-Gathering

Japan seems to have established and be able to maintain a well-organized network for scientific and technological information-gathering. The committee recommends that the United States strengthen its effort along those lines with many different approaches. They could include more vigorous examination of technology assessment in Japan, more support for exchange-scholar and exchange-student programs, collaborative research programs, and establishment of a centralized coordinating and networking infrastructure with its own objectives, policy, and strategy. For bioprocess engineering, particular emphasis should be placed on tracking developments in bioprocess technology for manufacture of new bioproducts.

3.1.4 Summary of Comparison With United States

A recent mission through the Japan Technology Evaluation Center was made by U.S. panel members to assess the status of bioprocess engineering and biotechnology in Japan. Present and likely future Japan-U.S. comparisons related to the various elements of bioprocessing are summarized in Figure 3.1. The present status is indicated by "+," "O," and "-" to show that Japan is ahead of, even with, or behind the United States, respectively. The future trends are indicated by arrows pointing up (Japan's capabilities will surpass those of the United States), horizontal (capabilities will be similar), and pointing down (U.S. capabilities will exceed Japan's).

3.2 BIOPROCESS ENGINEERING IN GERMANY AND EUROPE

One of the charges of the committee was to benchmark bioprocess engineering in Europe. The committee emphasized Germany, because some of the members were familiar with Germany's programs; and two members visited Germany in the fall of 1991. However, other countries in Europe are also very strong in bioprocessing and biotechnology, including Switzerland, the United Kingdom, France, Sweden, Denmark, Norway, Italy, France, Austria, and the Netherlands.

3.2.1 Education

The education of bioprocess engineers in Germany is usually accomplished mainly within a few departments of chemical engineering or chemical technology. Scientific departments, such as those in microbiology and biochemistry, also contribute to the education of bioprocess engineers in Germany. Germany has traditionally directed more of its efforts in education to the chemical and biological sciences than to engineering.

The major difference in educational training in universities between Germany and the United States lies in the German system in which a single professor is in charge of a department. Assistant, associate, or full professors who might be part of the educational program in a department in the United States and add diversity to educational programs are not prevalent in Germany. Most of the bioprocess engineers are trained at the graduate level, as is the case in the United States, Japan, and other western European countries.

Molecular Biology	Product Discovery		Genetics	
	Now	Future	Now	Future
	−	↘ (Japan will be behind)	−	↘ (Japan will be behind)

Microbiology	Screening		Strain Development		Fermentation Technology	
	Now	Future	Now	Future	Now	Future
	+	↗ (Japan ahead)	+	↗	+	↗

Upstream Processing	Process Development		Engineering Science		Monitoring, Biosensors, Control		Bioreactor Scaleup	
	Now	Future	Now	Future	Now	Future	Now	Future
	O	➤ (similar)	O	➤	+	↗	−	➤

Downstream Processing	Solid-Liquid Separation		Cell Disruption		Membrane Technology		Affinity Chromatography	
	Now	Future	Now	Future	Now	Future	Now	Future
	O	➤	O	➤	−	➤	−	↘
	Ion-Exchange Chromatography		Size-Exclusion Chromatography		High Pressure Liquid Chromatography		Protein Folding	
	Now	Future	Now	Future	Now	Future	Now	Future
	−	➤	O	➤	O	➤	−	➤

Biocatalysis	Enzyme Discovery		Enzyme Science		Enzyme Engineering		Industrial Implementation	
	Now	Future	Now	Future	Now	Future	Now	Future
	+	↗	−	↗	+	↗	+	↗

Other Manufacturing Issues	Containment		Good Manufacturing Practices		Technology Transfer	
	Now	Future	Now	Future	Now	Future
	−	↘	−	↘	+	↗

Educational Status	Basic Training		Applied Training		Engineering vs. Science		Faculty Biotechnology Knowledge	
	Now	Future	Now	Future	Now	Future	Now	Future
	−	↘	+	↗	−	↘	−	➤

University, Government, Industry Relations	University-Industry		University-Government		Government-Industry		Overall	
	Now	Future	Now	Future	Now	Future	Now	Future
	+	↗	+	↗	+	↗	+	↗

Present: + = Japan ahead, O = even, − = Japan behind. Future: ↗ = Japan will be ahead, ➤ = two countries will be similar, ↘ = Japan will be behind.

FIGURE 3.1   Comparison of Japanese with U.S. bioprocessing.

3.2.2 University Research in Germany

German research universities have components of applied and basic research. That differs from the United States, where the fundamental principles are stressed at the graduate level. Research programs in Germany encompass practical and applied research, in addition to fundamental engineering research. Another major difference between the United States and Germany, in terms of university research, is the Fraunhofer institutes, or bioprocess engineering centers, supported by the German states and placed throughout Germany. Those institutes, directly associated with universities, address both fundamental bioprocess engineering and process development. Most of the Fraunhofer institutes are well equipped with bioprocess equipment for laboratory to pilot-scale operations. An important element of a Fraunhofer institute is the close association between the state government, the university, and industry. Much of the research and development at the Fraunhofer institutes are conducted to meet industrial needs, and it is possible to conduct research with confidentiality for a company in the confines of a Fraunhofer institute.

A number of national centers in bioprocess engineering deal with bioprocess engineering. The National Research Center for Biotechnology (GBF) at Braunschweig and the bioprocess-engineering center at Julich are examples. The national centers are associated with universities, and student training, as well as research, takes place there. Like the state Fraunhofer institutes, the federally supported centers house activities ranging from basic research and process development to scaleup with small-scale and pilot-scale equipment. Both the Fraunhofer institutes and the national centers often examine process development or process feasibility for industrial processing. Like the state-supported centers, GBF and Julich are able to collaborate with industry and perform research on behalf of industry with a high degree of confidentiality. The budgets of state and federal centers are provided for long periods. For example, the GBF in Braunschweig has been in existence for over 10 years.

Bioprocess-engineering research is conducted in the industrial sector. Many biotechnology companies in Germany focus heavily on the new biotechnology of recombinant DNA, as well as the traditional fermentation technology, such as that for antibiotics and large-volume chemical production. Industrial research in Germany is similar to that in the United States. A major difference between the United States and Germany is a gradual tendency for the production activities of major German companies to gravitate from Germany to the United States. For example, the acquisition of Miles Laboratory by Bayer has established a strong foothold in recombinant-DNA technology for Germany within the United States. Other examples are BASF and Henkel. The environmental and "green" policies have made it more difficult for German industry to enter the recombinant market with manufacturing based in Germany. Therefore, much of the bioprocess engineering in the major German multinational companies will be developed in part in the United States, and much of the manufacturing for the companies will also take place on U.S. soil.

Figure 3.2 summarizes the opinions of the committee on the present and future states of European biotechnology processing relative to that in the United States. The comparison includes Germany, Switzerland, Austria, United Kingdom, France, Scandinavia, Italy, and the Netherlands and represents an aggregate estimate of status and direction based on committee members' experience and knowledge of European biotechnology. Given some of the excellent biotechnology products and services available from these countries and the upcoming economic unification of Europe, we recommend a separate study on bioprocessing in Europe.

Molecular Biology	Product Discovery		Genetics	
	Now	Future	Now	Future
	-	↘	-	↘

Microbiology	Screening	Strain Development	Fermentation Technology
	O →	O →	O →

Upstream Processing	Process Development	Engineering Science	Monitoring, Biosensors, Control	Bioreactor Scaleup
	O →	O →	O →	O →

Downstream Processing	Solid-Liquid Separation	Cell Disruption	Membrane Technology	Affinity Chromatography
	O →	+ →	- →	O →
	Ion-Exchange Chromatography	Size-Exclusion Chromatography	High Pressure Liquid Chromatography	Protein Folding
	O →	+ ↗	O ↘	O →

Biocatalysis	Enzyme Discovery	Enzyme Science	Enzyme Engineering	Industrial Implementation
	O ↗	O →	O ↗	+ ↗

Other Manufacturing Issues	Containment	Good Manufacturing Practices	Patents	Technology Transfer
	O →	O →	- →	+ ↗

Educational Status	Basic Training	Applied Training	Engineering vs. Science	Faculty Biotechnology Knowledge
	- →	+ →	+ →	- →

University, Government, Industry Relations	University-Industry	University-Government	Government-Industry	Overall
	+ →	O ↘	O →	O →

Present: + = Europe ahead, O = even, – = Europe behind. Future: ↗ = Europe will be ahead, → = Europe and U. S. will be similar, ↘ = Europe will be behind.

FIGURE 3.2 Comparison of European with U.S. bioprocessing.

3.3 BIOPROCESS ENGINEERING IN UNITED STATES

The current status of bioprocessing in the United States provides a useful reference against which the estimated magnitude of future developments can be gauged. The commercialization of products that use or are obtained from the new biotechnology is relatively recent. Within only the last 5 years, sales of biotechnology products have grown from approximately $100 million a year to $4 billion a year. Sales in 10 years are expected to be 10 times today's. Consequently, the committee undertook to estimate the current status of U.S. bioprocessing and particularly bioprocess engineering to provide a reference point for estimating future needs.

3.3.1 Education and Training

Bioprocess-engineering faculty in U.S. departments of chemical, biochemical, agricultural, civil, and mechanical engineering in 1990 numbered about 250. About 200 of them are in chemical-engineering departments

(ACS, 1991). In addition, faculty involved in bioprocess development in allied fields—including biology, applied microbiology, molecular biology, medicinal chemistry, and industrial pharmacy—are estimated to number about 50-75. On that basis, the total current U.S. faculty in bioprocess engineering and bioprocess technology number upwards of 300 engineering and science faculty. Those faculties are estimated to be able to award about 150-200 master's degrees and doctorates in biotechnology processing and bioprocess engineering a year in the coming 10 years. The numbers just cited do not include faculty involved in graduate training in departments of food science, biomedical engineering, or electrical engineering.

The education and training of bioprocess engineers in the United States is carried out primarily on the graduate level, although many engineering seniors have had one or two courses in introductory bioprocessing or biochemical engineering. According to sales of major textbooks used in the relevant courses and general chemical-engineering enrollment figures, it can be estimated that about 10,000 engineering seniors and graduate students have had at least an introductory exposure to bioprocess engineering during the last 15 years. Of those students, the committee estimates that less than 10% ultimately pursued graduate (M.S. or Ph.D.) or other specialized studies related to bioprocess engineering since 1977.

About 3,000 bioprocess-development technologists and engineers (B.S., M.S., and Ph.D.) are employed in an industry that uses new biotechnology for the manufacture of biotherapeutics and intermediate-value products, including enzymes, and for waste treatment (principally through bioremediation). They are probably increasing by fewer than 180 per year. The aggregate growth of bioprocess engineers over the next 10 years, if sustained at current levels, would be 75%. The projected growth rate of the industry is estimated to be 1,000%. There is a clear disparity between the need for manufacturing capability and the engineers and technologists who are likely to be prepared to provide the capability.

3.3.2 Government Initiative and Support

The biotechnology initiative of the National Science Foundation defines bioprocessing as encompassing the spectrum of events that produce a substance of biological origin. Bioprocessing research is defined in more detail as including:

• A fundamental understanding of the formation of the product.
• Knowledge of separation and purification methods.
• Process monitoring and control, systems analysis, and integration of upstream and downstream processing.

Vignette 5

Bioprocess-Industry Needs: A Moving Target

Staffing needs in bioprocessing are a moving target in an industry in which 1 year represents 10%-20% of the life history of many companies. As little as a year ago (in the first part of 1991), a survey of small to medium U.S. biotechnology firms (of which there are now about 750) showed that government regulation and availability of scientific personnel were causing substantial concern; the availability of fermentation and bioprocessing expertise was not viewed as an immediate issue (Dibner, 1991).

The average U.S. biotechnology firm appears to have 98 employees, of whom 19 are in basic research, 17 in product development, 22 in production, and nine in marketing. Genentech had an estimated 780 production employees among a total of 2,020. In about 750 biotechnology firms, production jobs are likely to account for 12,000 employees. When the biotechnology-related production employees of larger corporations are added, the number is likely closer to 17,000. The survey (by Dibner, 1991) was taken in the spring of 1991, when few biotechnology firms were showing profitability (Thayer, 1991). It targeted primarily the biopharmaceutical sector and did not include the substantial impact of agriculture-derived and value-added products or the engineering requirements for second-generation biotechnology products, which will need to be produced at higher volume and lower cost than the first-generation ones (see Vignettes 3 and 4).

Rapid changes in perceived needs are inevitable in the rapidly growing biotechnology industry (Clemmitt, 1992). Hence, many issues will need to be addressed quickly. Abelson (1992), in assessing the 1991 Office of Technology Assessment (OTA) report, suggests that "the United States will remain a substantial factor in the commercialization of biotechnology. However, a dominant role is being frittered away." The committee believes that basic planning and allocation of resources to ensure availability of a well-trained group of bachelor's-, master's-, and doctorate-level bioprocess engineers would moderate such a trend. Therefore, expanded activities in training must be an immediate concern, given the 2- to 5-year interval that passes between initiation and completion of training.

- A basic understanding of molecular, genetic, metabolic, and cellular function and regulation in culture and bioreactors.
- Development of new approaches to bioprocessing.

The committee concurs with the assessment given in the Federal Coordinating Council for Science, Engineering, and Technology (FCCSET) report (1992, p. 44): "Manufacturing/Bioprocessing is an area in which biotechnology offers vast potential rewards. The total federal investment of $99 million in FY 1992 is small in proportion to its potential. . . . It is a priority of the Biotechnology Research Initiative to significantly enhance research in this area." The funding in bioprocessing will need to be gradually increased, as the industry grows past the level reflected by the funding requested for FY 1993. Assuming a 4% average inflation rate and a 6% growth rate, an increase of around 10% per year would be needed. That is equivalent to growth of the programs from $43 million per year (budget request for FY 1993) to $110 million per year 10 years from now.

The application of bioprocess-engineering principles to separation and purification methods has high priority, because separation and purification costs are often 50% or more of processing costs. Ancillary contributions to improving productivity and product quality and decreasing costs of purification will come from improvements in process-monitoring control, systems analysis, and integration of upstream and downstream processing. Research and development on those topics, if structured to address techniques that affect industrial manufacturing processes, will complement fundamental research in developing needed biological and biochemical tools for assays and downstream processing discussed in Chapter 4. Combined with improvements in the basic understanding of the productivity of cell culture and bioreactors, research on downstream processing, systems integration, and process monitoring could contribute substantially to the manufacturing technology base and at the same time provide training for bioprocess engineers who would work in the industry. Given the opportunities for training and having a major impact in the industry, strong government support of research in bioprocess engineering for separation and purification is appropriate, if not critical, to U.S. bioprocessing industries. Research in purification and separation will be most effective if structured to foster cross-disciplinary approaches that seek the application of biological, chemical, and biochemical principles. To that end, sustained increases in NSF resources for bioprocessing programs need to be maintained.

Another key challenge for the protein-pharmaceutical industry is the effective analysis of the final product and product intermediates. The product must consistently be shown to be identical with the product that was demonstrated as safe and effective in human clinical trials. Proteins are large complex molecules, so the task is often difficult and expensive and might be rate-limiting in a product-development effort. The relationship between analytical characterization of a molecule and clinical performance is particularly difficult to determine. For example, it is not now possible to know whether a molecular form will be immunogenic or even to know what characteristic of the molecule determines its immunogenicity. Such limitations also often make it expensive and difficult to improve a process after a product is approved for marketing.

To address that need, it is recommended that greater cooperation be established between the protein-pharmaceutical industry and FDA. Conveyance of industrial viewpoints to FDA can now occur through committees of the Pharmaceutical Manufacturers Association and the Industrial Bio-

technology Association. Similar arrangements should be established to foster and support active communication and research aimed at developing
more effective and less expensive methods for protein-pharmaceutical characterization.

3.4 SUMMARY

The excellent basic-research program in the biological sciences has ben-
efited the U.S. biotechnology industry, as well as being a valuable source of
information for development of bioprocessing technology in both Japan and
Europe. The committee feels that, in product discovery and genetics, as
related to molecular biology, both Europe and Japan are behind the United
States and will continue to be behind for the foreseeable future. That
reflects, in part, the important long-term commitment that has been made to
the basic sciences (at $3.0 billion per year) as they are related to the new
biotechnology through government-funded research programs (FCCSET,
1992). In many of the other fields illustrated in Figure 3.2, Europe and the
United States are approximately equivalent, and their competitive positions
are likely to be maintained for the foreseeable future. Japan, in contrast, is
ahead and will remain ahead for the foreseeable future (see Figure 3.1) in
monitoring and control, biocatalysis, applied training, and university-gov-
ernment-industry relations.

If the U.S. biotechnology industry is to remain competitive with those of
Japan and Europe, a major commitment is needed to develop engineering
staffing in bioprocess development and to support manufacturing technologies. We recommend a major commitment to developing the manpower
base through funding of research programs in universities, continuing-
education programs, and research directed toward industrial problems
(applied-engineering research). New resources must be provided to increase the infrastructure for bioprocess engineering and biotechnology in
this context.

3.5 REFERENCES

Abelson, P. H. 1992. Biotechnology in a global economy [editorial]. Science 255(5043):381.
ACS (American Chemical Society). 1991. Directory of Graduate Research. Washington,
    D.C.: American Chemical Society. 1561 pp..
Clemmitt, M. 1992. Maturing biotech firms face new challenges. The Scientist 6(8):1.
Dibner, M. D. 1991. Manpower—Present and Future—in Bioprocessing: Perspectives from
    the Biotechnology Industry, Report to the Committee on Bioprocess Engineering, Board
    on Biology, National Research Council, Washington, D.C.
FCCSET (Federal Coordinating Council for Science, Engineering, and Technology). 1992.
    Biotechnology for the 21st Century. A Report by the FCCSET Committee on Life
    Sciences and Health, Office of Science and Technology Policy, Executive Office of the
    President, Washington, D.C.

JTEC (Japanese Technology Evaluation Center). 1992. Bioprocess Engineering in Japan. NTIS Report No. PB92-100213, Washington, D.C.

OTA (Office of Technology Assessment). 1991. Biotechnology in a Global Economy, B. Brown, ed. Office of Technology Assessment, U.S. Congress, Report No. OTA-BA-494. Washington, D.C.: U.S. Government Printing Office.

Thayer, A. M. 1991. Revenues grow for biotech firms but few show profitability. Chem. Eng. News 69(36):17-18.

4

Current Bioprocess Technology, Products, and Opportunities

Products and services that depend on bioprocessing can be grouped broadly into

• *Biopharmaceuticals.* Therapeutic proteins, polysaccharides, vaccines, and diagnostics.
• *Specialty products and industrial chemicals.* Antibiotics, value-added food and agricultural products, and fuels, chemicals, and fiber from renewable resources.
• *Environmental-management aids.* Bioprocessing products and services used to control or remediate toxic wastes.

This chapter reviews the status of bioprocessing for manufacture of products in categories that are relevant for the next 10 years. Much of the relevant background is derived from an Office of Technology Assessment report, *Biotechnology in a Global Economy* (OTA, 1991).

4.1 BIOPHARMACEUTICALS

The success of biotechnology is seen in the impact of new products and processes. The products include biotherapeutics, specialty chemicals, and reagents, such as diagnostics, biochemicals for research, and enzymes for the food and consumer markets. The purpose of this section is to examine the state of bioprocessing of biopharmaceuticals, including the status of current research and the needs and opportunities for innovation in bioprocessing for manufacturing of biotherapeutic products. *Biotherapeutics* include therapeutic proteins, vaccines, therapeutic polysaccharides, diagnostics, and low-molecular-weight pharmaceutical chemicals.

The development of recombinant-DNA and hybridoma technologies has revolutionized the array of pharmaceutical products available. Unlike traditional therapeutics, the pharmaceuticals produced by the new technologies are primarily protein products; they include insulin, growth hormone, α-interferon, OKT-3 monoclonal antibody, tissue plasminogen activator, hepatitis vaccine, and erythropoietin. With the availability of large amounts of those products, new clinical applications are being discovered. For example, it has been discovered that growth hormone is effective in wound healing, in addition to the treatment of pituitary dwarfism.

Although regulatory requirements for safety and efficacy lead to long delays in the approval of biotherapeutic products for sale, 15 products had been approved in the United States by the end of 1991 (Table 4.1). Estimates of annual sales range from $3 to 5 billion for 1991 and constitute about 7-10% of total U.S. pharmaceutical sales. Most noteworthy are the increase of more than 10% in annual sales of existing biotherapeutics and the large number (158) of products in the clinical-trial pipeline (Tables 4.2 and 4.3) with an expectation that a substantial number will be approved for therapeutic use. Clearly, biotherapeutics have an important role in improving human health care. The important question to be addressed here is: What are the technological needs in the next decade related to facilitating manufacturing and commercialization of products evolving from biotechnology? To address that question, we first need to examine and understand the current state of bioprocessing.

4.1.1 Proteins from Recombinant Microorganisms

Extensive research on eukaryotic gene expression in bacteria, yeasts, plants, insects, and mammals has resulted in many options for producing proteins in recombinant hosts. In spite of the numerous options, most of the products manufactured today are made either in recombinant *E. coli* or in animal cells, i.e., Chinese hamster ovary (CHO) cells or hybridoma cells.

E. coli is the microbial system of choice for the expression of heterologous proteins. No other microorganism is used to produce so large a number of products at high level. Typical levels of foreign protein expressed represent 10-30% of total cellular protein. Rapid progress in the development of *E. coli* as a host for foreign-gene expression is due mainly to *E. coli*'s having been the focus of intense study over the last 50 years in academic laboratories. The body of knowledge that has accumulated has facilitated the adaptation of this bacterium for foreign-protein expression. Sophisticated cloning vectors, tools for regulated gene expression, and knowledge about the process of protein secretion and the physiology of growth were available in *E. coli*, and it became the logical choice for heterologous-gene expression.

Table 4.1 Approved Biotechnology Drugs and Vaccines

Product name	Company	Indication	U.S. Approval	Revenues[a] 1989	Revenues[a] 1990
Epogen ™[b] Epoetin Alfa	Amgen Thousand Oaks, CA	Dialysis anemia	June 1989	95	300
Neupogen[b] Granulocyte colony stimulating factor G-CSF	Amgen Thousand Oaks, CA	Chemotherapy effects	February 1991	NA	NA
Humatrope ®[b] Somatotropin rDNA origin for injection	Eli Lilly Indianapolis, IN	Human growth hormone deficiency in children	March 1987	40	50
Humulin ® Human insulin rDNA origin	Eli Lilly Indianapolis, IN	Diabetes	October 1982	200	250
Actimmune[b] Interferon gamma 1-b	Genentech San Francisco, CA	Infection/chronic granulomatous disease	December 1990	NA	NA
Activase ® Alteplase, rDNA origin	Genentech San Francisco, CA	Acute myocardial infarction	November 1987	175	200
Protropin ®[b] Somatrem for injection	Genentech San Francisco, CA	Human growth hormone deficiency in children	October 1985	100	120
Roferon ®-A[b] Interferon alfa-2a (recombinant/Roche)	Hoffmann-La Roche Nutley, NJ	Hairy cell leukemia AIDS-related Kaposi's sarcoma	June 1986 November 1988	40	60

Leukine[b] Granulocyte macrophage colony stimulating factor GM-CSF	Immunex Seattle, WA	Infection related to bone marrow transplant	March 1991	NA	NA
Recombivax HB ® Hepatitis B vaccine (recombinant MSD)	Merck Rahway, NJ	Hepatitis B prevention	July 1986	100	110
Orthoclone OKT ® 3 Muromonab CD3	Ortho Biotech Raritan, NJ	Kidney transplant rejection	June 1986	30	35
Procrit[b] Erythropoietin	Ortho Biotech Raritan, NJ	AIDS-related anemia Pre-dialysis anemia	December 1990	NA	NA
HibTiter ™ Haemophilus B conjugate vaccine	Praxis Biologics Rochester, NY	Haemophilus influenza type B	December 1988	10	30
Intron ® A[b] Interferon-alpha2b	Schering-Plough Madison, NJ	Hairy cell leukemia	June 1986	60	80
		Genital warts AIDS-related Kaposi's sarcoma	June 1988 November 1988		
		Hepatitis C	February 1991	NA	NA
Energix-B Hepatitis B vaccine (recombinant)	SmithKline Beecham Philadelphia, PA	Hepatitis B	September 1989	20	30

[a]Estimated U.S. revenues in millions of dollars.
[b]Orphan drug.
NA = not applicable.
SOURCE: OTA, 1991, p. 77.

Table 4.2   Conditions for Which Biotechnology-Derived Drugs are Under Development

AIDS and AIDS-related complex (ARC)
Chemotherapy effects
Leukemia
Aplastic anemia
Cancer
Bone marrow transplant
Hematological neoplasms
Neutropenia
Myelodysplastic syndrome
Infectious diseases
Thermal injury
Reperfusion injury related to myocardial infarction and renal transplantation
Anemia secondary to kidney disease, AIDS, premature infants, chemotherapy, rheumatoid
    arthritis
Autologous transfusion
Hemophilia
Corneal transplants
Wound healing
Chronic soft tissue ulcers
Diabetes
Wasting syndromes
Nutritional and growth disorders
Venous stasis
Turner's stasis
Burns
Venereal warts
Herpes simplex 2
Hepatitis-B, non-A non-B hepatitis
Hypertension
Platelet deficiencies
Septic shock
Pseudomonas infections
Heart and liver transplant rejection
Malaria
Need for cervical ripening to facilitate childbirth
Myocardial infarction
Deep vein thrombosis
Acute stroke
Pulmonary embolism

SOURCE: OTA, 1991, p. 77.

Table 4.3 Additional Indications for Approved Drugs

Drug	Approved Indications	Additional Indications
EPO	Dialysis anemia, AIDS-related anemia	Autologous transfusion, prematurity, rheumatoid arthritis, chemotherapy
Tissue plasminogen activator	Acute myocardial infarction	Deep vein thrombosis, acute stroke, pulmonary embolism
Interferon α-2a	Hairy cell leukemia, AIDS-related Kaposi's sarcoma, hepatitis C	Cancer, infectious disease, genital herpes, colorectal cancer, chronic and acute hepatitis B, chronic myelogenous leukemia, gastric malignancies, HIV-positive ARC, AIDS
Interferon γ-2b	Hairy cell leukemia, genital warts, AIDS-related Kaposi's sarcoma	Genital herpes, superficial bladder cancer, basal cell carcinoma, chronic and acute hepatitis B, non-A non-B hepatitis, delta hepatitis, chronic myelogenous leukemia, HIV

SOURCE: OTA, 1991, p. 77.

Expression in *E. coli* can now be designed for either intracellular accumulation of the heterologous protein in the cytoplasmic space or translocation of the protein across the cytoplasmic membrane from the cytoplasmic space into the periplasmic space. After translocation, the protein can accumulate within the periplasmic space or might be released to the surrounding medium. If the protein is secreted and accumulated within the cytoplasmic space, it normally aggregates into large inclusion bodies visible with a light microscope. These must be isolated, solubilized, and folded to obtain an active molecule. Isolation and solubilization are routine, but folding to an active form is difficult with present technology. Intracellular accumulation often has the additional disadvantage of producing a substance with an extra amino acid on the N terminus of the protein.

Several signal sequences are now available to drive the secretion of eukaryotic proteins across the bacterial cytoplasmic membrane. Occasionally, that results in the formation of properly folded, bioactive proteins. More often, however, the secreted proteins also accumulate as aggregates in the periplasmic space; again, it is necessary to isolate, solubilize, and fold the proteins to their proper conformation. Perhaps the most exciting application for secretion of proteins from *E. coli* has been the demonstrated ability to produce active fab fragments (the binding portion of antibodies) directly; these will have a variety of uses as assay reagents, immunoaffinity ligands, and therapeutics.

For both intracellular and secreted eukaryotic proteins, proteolytic degradation in *E. coli* is a problem. Several approaches have been taken to reduce undesirable proteolysis, including the expression of fusion proteins and the elimination of specific proteases by host-cell mutation. The latter approach has been useful, but continued removal of proteases can be expected to affect general cellular metabolism adversely. Mistranslation has also been an occasional problem, but published technology now exists to minimize it.

In summary, *E. coli* expression of eukaryotic proteins has been an important "workhorse" for the production of rDNA proteins. The cells grow and express rDNA proteins rapidly and in high quantities. They also are easily modified genetically and generally require inexpensive growth media. However, the system is often limited by its inability to produce intact, properly folded proteins and by a limited ability to yield posttranslational modifications, such as glycosylation and specific proteolytic modification. Nonetheless, the system has enabled the commercialization of such products as human insulin, human growth hormone, human α-interferon, and human γ-interferon.

4.1.2 Inclusion Bodies

High levels of protein synthesis have been obtained with several intracellular expression systems, particularly in *E. coli*. High expression of a foreign protein in the cytoplasm of *E. coli* often results in the accumulation of nonnative aggregates called inclusion bodies. Isolation of inclusion bodies by centrifugation has become an important first step in the purification and recovery of recombinant proteins.

Extensive protein-chemistry studies have revealed substantial fundamental information on the mechanism of inclusion-body formation. Various solubilization agents have been defined (strong chaotropes, detergents, and organic solvents) for use in recovery of active proteins; the process requires unfolding the protein with strong denaturants and refolding to an active monomer.

Studies of the refolding of denatured proteins both in vitro and in vivo indicate that aggregates derive from specific partially folded intermediates and not from mature native or fully unfolded proteins (Mitraki and King, 1989). Those discoveries focused attention on the properties of intermediates (as distinct from native states) and the factors interacting with them, such as the intracellular cytoplasmic environment, cofactors, and molecular chaperones.

Molecular chaperones were first identified as host proteins needed for phage morphogenesis and have recently been identified as heat-shock proteins (Goloubinoff et al., 1989). In a recent review (Pelham, 1986), it was proposed that heat-shock proteins can act as molecular chaperones and prevent aggregation by binding to hydrophobic regions of partially unfolded polypeptide chains. On the basis of those fundamental discoveries, studies are under way to mimic the mechanics of mammalian protein synthesis (compartmentation, interprotein interactions, and posttranslational modifications) in bacteria. With rational selection of the characteristics necessary for correct maturation, it might be possible to direct the fate of the intermediates toward the native conformation. Alternatively, it might be possible to use molecular chaperones to repair and disaggregate proteins outside the cell before releasing them for refolding to the active monomer.

4.1.3 Mammalian Host Systems

Production of heterologous proteins by mammalian cells has usually used CHO cells or hybridoma cells. Initially, hybridoma cells were the only hosts used for antibody production. More recently, CHO cells and mouse myeloma cells have also been used. CHO cells are generally able to produce bioactive mammalian proteins that are glycosylated and properly folded. As yet, the system is often not able to effect specific proteolytic maturation, except to remove the secretion-signal sequence.

Although bioactive molecules are usually formed by CHO cells, the product is a mixture of many subforms that differ in degree of glycosylation, electrostatic charge, the presence of proteolytic clips, and other possible modifications. The modifications do not necessarily compromise the potency or safety of the product, but it is essential that the process be carefully controlled to ensure that the same profile of molecular variants is produced from each batch.

Mammalian cells have the advantage of being able to produce complex, bioactive molecules. However, they grow and express proteins at approximately one-twentieth the rate of *E. coli*. That has the effect of increasing capital and labor costs for protein production. The cells also require expensive media (although efforts are under way to reduce these costs) and have additional, although tractable, regulatory and safety concerns, such as concern about undetected viral contamination. In spite of those limitations, CHO-cell production of biopharmaceuticals is an established and important technology that has enabled the delivery of such important therapeutics as tissue plasminogen activator and erythropoietin.

4.1.4 Other Hosts for Heterologous Gene Expression

Several new systems for the production of heterologous proteins are under development. They include such new bacterial systems as *Bacillus* and *Streptomyces*, the filamentous fungi, insect cell lines of *Drosophila*, and systems that rely on the baculovirus expression system, *Xenopus* oocytes, and yeast. Although none of these is as developed or has been studied as extensively as *E. coli*, each has advantages and disadvantages. In *Bacillus*, for example, strains that lack most of the usual proteases have been generated. *Streptomyces* does not compete with *E. coli* in level of expression, but is useful for making small quantities of soluble, nature-identical product. Filamentous fungi, such as *Neurospora crassa* and *Aspergillus nidulans*, can secrete copious quantities of protein and have long been used in the pharmaceutical industry to make natural products. Yeast has been used to produce rDNA proteins, such as IGF-1 and human serum albumin; in spite of substantial effort, it has not been used as extensively as *E. coli* or CHO cells.

4.1.5 Isolation and Purification

Isolation generally denotes the separation of the product from the bulk of the producing organism. The disposition and state of the expressed protein affect the isolation procedure. For mammalian cells and some *E. coli*, *Streptomyces*, *Bacillus*, and yeast products, the protein is released from the cell into the surrounding medium, and isolation is effected by a solid-liquid separation step, usually centrifugation or microfiltration or ultrafiltration. If the product has aggregated either in the cytoplasmic or periplasmic space, isolation is more involved. Generally, the cell is first lysed by mechanical, chemical, or enzymatic treatment (or a combination). In some cases, the more dense aggregate can be separated by centrifugation from most of the soluble and insoluble cell components; in other cases, the aggregate is first solubilized while still in the soluble protein mixture.

Purification of the protein is a critical and often expensive part of the process. It might account for 50% or more of the total production cost. Purification has several objectives: to remove contaminating components from the host organism, i.e., other proteins, DNA, and lipids; to separate the desired protein (or family of proteins) from undesired variants of the desired protein; to remove and avoid the introduction of endotoxin; to inactivate viruses; to obtain required yields at acceptable cost; to avoid chemical or biochemical modification of the protein; and to make the process consistent and reliable. In some cases, the first and additional objective is to fold the protein into its desired conformation.

Much of the accumulated knowledge about protein purification is the property of individual companies. However, the available information suggests a general consistency in the type and order of process steps. The most common individual operations are centrifugation, filtration, membrane separation, adsorption separation, and chromatography.

Regulatory and safety concerns have combined with the desire for stable liquid formulations to motivate the removal of host-organism proteins to a maximal degree. Measurement of those contaminants requires sophisticated assays capable of detecting a spectrum of possible contaminants at a few parts per million of the product protein. The presence of undesired variants of the target protein has motivated the development of techniques to detect and separate (on a large scale) proteins modified at one of several hundred amino acids.

The difficulty of separation can often be decreased by changing the organism or culture conditions to produce a more uniform protein. However, it is still necessary to combine a series of purification steps each of which separates according to a different principle. Ultrafiltration steps are often used between separation steps to concentrate the protein solution or to make the buffer solution compatible with the next separation step. The final steps are designed to place the purified protein in the solution used for the product form.

The complexity of the individual purification steps and the need to be able to integrate them into a manufacturing system translate into a major opportunity for bioprocessing engineering as the process moves from the bench to the plant. Research and development in purification, scaleup integration, and system design will continue to have high priority.

4.1.6 Protein Engineering

Advances in molecular biology have provided researchers with the opportunity to develop increasingly rational approaches to the design of therapeutic drugs. This technology, when used with computer-assisted molecular modeling, is called *protein engineering*. Protein engineering combines many techniques, including gene cloning, site-directed mutagenesis, protein expression, structural characterization of the product, and bioactivity analyses; it can be used to modify the primary sequence of a protein at selected sites to improve stability, pharmacokinetics, bioactivity, and serum half-life.

A second application of protein engineering is the design of hybrid proteins that contain regions that aid separation and purification. That is achieved by introducing, next to the structural gene for the desired product, a DNA sequence that encodes for a specific polypeptide "tail." The tails can be inserted at the N or C terminal of the protein to yield a fusion protein with special properties that facilitate separation. Such genetic modifications can be designed to take advantage of affinity, ion-exchange, hydrophobic, metal-chelate, and covalent separations. Examples of affinity tails and the corresponding ligands are given in Table 4.4. The special properties of fusion proteins allow crude microbial extracts to be passed over an adsorbent that binds specifically to the tail, so that the desired product is retained and contaminants pass through. After elution and treatment to remove the tail, the product is purified further by standard methods, such as size-exclusion chromatography or high-performance liquid chromatography (HPLC).

4.1.7 Glycobiology

Recent studies of receptor biology have resulted in fundamental discoveries about the role of complex oligosaccharides in disease, in modulation of protein function, and as anchors for integral membrane glycoproteins. As additional glycoproteins are identified and cloned, there is an increasing need for more effective chromatographic methods, production systems that mimic mammalian glycosylation patterns, and fast, reproducible analytical methods to minimize microheterogeneity during manufacture.

Variability in oligosaccharide biosynthesis has been found to be an important source of heterogeneity for glycoproteins produced by eukaryotic cells (Marino, 1989). Glycoprotein oligosaccharides are covalently attached to proteins through the amino acid serine (O-linked) or asparagine (N-linked). If a selected carbohydrate type and site are required for bioactivity of a candidate glycoprotein, the expression system must be carefully selected.

Table 4.4 Examples of Affinity Tails

Tail	Ligand
β-Galactosidase	β-D-Thiogalactoside
Protein A	IgG
Protein G	IgG or albumin
Chloramphenicol acetyl transferase	β-Amino chloramphenicol
Glutathione S-transferase	Glutathione
Avidin	Biotin
Streptavidin	Biotin
Arginine	Anion exchange
Glutamate or aspartame	Cation exchange
Cysteine	Thiol
Histidine	Ni^{2+}, Cu^{2+}, Zn^{2+}
Phenylalanine	Phenyl
Antigenic peptide	Monoclonal antibody

SOURCE: Derived from Hammond et al., 1991.

Bacterial systems cannot glycosylate, many yeast species hyperglycosylate, and glycosylation in mammalian cells has been shown to be specific to tissue and cell type.

Future challenges in bioprocess development will parallel research in glycoprotein chemistry. The development of appropriate process controls, analytical methods, and quality-control specifications to control lot-to-lot consistency will be complicated by the inherent microheterogeneity of glycoproteins.

4.1.8 Metabolic Engineering

A powerful new approach to product development is the creative application of fermentation technology and molecular biology for "metabolic engineering." Examples of metabolic engineering for heterologous-protein production include deletion of proteases that eliminate product and production of factors that facilitate product maturation and secretion. For protein production on an industrial scale, metabolic engineering could be useful in shifting metabolic flow toward a desired product, creating arrays of enzymatic activities for synthesis of novel structures, and accelerating rate-limiting steps (Bailey, 1991). Metabolic engineering has recently been used to increase the efficiency of nutrient assimilation (increasing the growth rate), improve the efficiency of ATP production (decreasing nutrient demands), and reduce the production of inhibitory end products (increasing final cell densities).

Central to molecular modification of multigene pathways, such as those involved in antibiotic production, is the development of new vectors and transformation procedures and other tools of molecular biology. Another important discovery in metabolic engineering is the isolation of positive-control genes that regulate production of secondary metabolites. Positive regulators have been found in biosynthetic gene clusters for actinorhodin, bialophos, streptomycin, and undecylprodigiosin, all of which are *Streptomyces* products (Bailey, 1991).

An example of the feasibility of introducing new biosynthetic capabilities into industrial microorganisms by combining fungal and bacterial genes is given by Isogai et al. (1991). Genes encoding the converting enzymes D-amino acid oxidase and cephalosporin acylase were cloned from *Fusarium* and *Pseudomonas*, respectively, into the fungus *Acremonium chrysogenum*. Expression of this "artificial" antibiotic biosynthetic pathway was confirmed by analysis of transformants that synthesized and secreted detectable amounts of 7-aminocephalosporanic acid.

In addition to classical mutation, new tools have become available for genetic manipulation of important producers of natural products, such as *Streptomyces*. The ability to clone and manipulate biosynthetic genes for antibiotic production, regulatory genes for improved synthesis, and genes from primary metabolic pathways that contribute to secondary biosynthetic pathways can facilitate construction of strains that have substantially altered metabolic properties. In addition, the cloning of heterologous genes into bacterial hosts has generated strains that can produce compounds that are foreign and even deleterious to cell physiology.

4.1.9 Polymerase Chain Reaction

Within the last 5 years, the polymerase chain reaction (PCR) has transformed the way DNA analysis is performed. The ability to amplify, as well as modify, a specific target DNA sequence with a template in a simple automated procedure has facilitated many tasks in molecular biology (e.g., cloning and sequencing) and opened up new fields for study.

The utility of PCR instrumentation was greatly expanded with the discovery of *Taq* DNA polymerase, a thermostable polymerase from *Thermus aquaticus* (Erlich et al., 1991). The use of the enzyme allowed the development of an automated thermal-cycling device for carrying out the amplification reaction in a single tube.

New approaches to improve specificity have also been developed with an approach called hot start. Amplification of nontarget sequences can be minimized by manual addition of an essential reagent (such as DNA polymerase or primers) to the reaction tube once it has reached a high temperature.

As advances in reagents and procedures continue to develop, researchers will apply PCR to an increasing number of problems in DNA analysis.

4.1.10 Monoclonal Antibodies and Antibody Engineering

The use of monoclonal antibodies has provided general access to homogeneous antibodies of prescribed specificity. Recent advances have dramatically altered monoclonal-antibody therapy. The advances include definition of cell-surface structures on abnormal cells as targets, development of genetic-engineering approaches for creating more effective agents, and development of techniques for arming monoclonal antibodies with radionuclides, toxins, or cytotoxic drugs to increase effector function (Waldmann, 1991).

The expression of antibody fragments in *E. coli* brings the arsenal of techniques of bacterial gene technology to antibodies. By systematically varying experimental characteristics, Buchner and Rudolph (1991) successfully produced Fab fragments at yields of up to 40% of the recombinant protein. Similarly, Davis et al. (1991) recently constructed a gene coding for a single-chain antibody and expressed functional, antigen-binding proteins in eukaryotic cells. Those advances are important for the development of antibodies as therapeutic products and as reagents.

4.1.11 Transgenic Animals

Transgenic animals are being developed for a wide variety of applications. In the near future, transgenic animals will be used increasingly in safety evaluation of new pharmaceuticals and accelerating their regulatory approval.

The feasibility of producing human pharmaceutical proteins in the milk of transgenic livestock has been established. As an alternative to cell-culture systems, such livestock appear to be appealing because of high volumetric productivity, low operating costs, capability of posttranslation modification of proteins, and potential for expansion of the producing organism. Bioprocess engineers face numerous technical challenges in converting a transgenic mammary gland system into a commercial prototype for large-scale manufacture of high-market-volume proteins, including

• Purification techniques for obtaining high-purity proteins that must be recovered and fractionated from a complex mixture of fats, proteins, sugars, and ions, some of which are in colloidal form (see Ruettimann and Ladisch, 1987).
• Optimization of product stability during recovery.
• Instrumentation to characterize posttranslation modifications made by the mammary gland "bioreactor."
• Development of on-line sensors to monitor changes in bioactivity of products during purification.
• Bioseparations of milk proteins.

In the longer term, transgenic animals might provide a source of tissues and organs for use in transplantation patients. Bioprocess engineering will be needed to design novel equipment to maintain, purify, and store the living tissues without affecting viability or graft response.

The hurdles to be surmounted in developing the necessary genetic tools for systematic pathway engineering are substantial, but basic research at the molecular level will continue to provide improved production strains and novel products, and continued interest in the fundamentals of bioprocessing of milk will help to define separation strategies for this complex biological fluid.

4.1.12 Product Formulation

Product formulation is an important and often overlooked part of bioprocess development. The product protein must be in a form that is stable, is convenient to use, and allows the drug to be delivered in the desired manner. Several of the initial rDNA products were in lyophilized form, which is relatively stable, but inconvenient. More recent biopharmaceuticals are in liquid form.

For lyophilized proteins, it is necessary to develop a carefully controlled freezing, drying, and capping procedure. The equipment is designed and tested to provide uniform conditions throughout the lyophilization chamber. For liquid formulations, production is simplified, but the selection of solution components is much more difficult. First, the adverse reactions must be identified for each protein product. Assays must then be developed with high sensitivity and accuracy to detect degradation products. Finally, conditions must be screened to find those which allow stable shelf-life of at least a year. Stability results might indicate that the protein needs to be purified further before formulation, for example, to remove a problem protease more completely .

4.1.13 Research Needs

The technology for bioprocesses in the manufacture of biopharmaceuticals covers a wide range of biological, chemical, and engineering disciplines. There are many research opportunities within this range, and the most important current industrial needs are listed below. The committee recommends that research funding be allocated to the topics listed here through a competitive-grants program. We believe that structuring the research in a manner that requires an industry-university or industry-government interaction would catalyze further research and help to promote approaches that are relevant to both the generic research and the future staffing needs of the developing industries. In addition to fostering the development of a fundamental engineering knowledge base and new technology oriented to bioprocess manufacturing, the program would help to train engineers who would be knowledgeable in bioprocessing as related to biopharmaceuticals and thereby be available for the staffing of both regulatory agencies and industrial facilities.

Biological and biochemical tools are needed and will require research to

• Develop tools for the expression, modification, and secretion of heterologous proteins from prokaryotic and eukaryotic cells. This includes the study of protein translocation and folding in *E. coli*, the stable insertion of foreign genes in mammalian cells, the elimination of protease production, and the development of tools for the use of new organisms for heterologous-gene expression.
• Modify proteins in vivo through protein engineering. This includes the study of chemical mechanisms and environmental influences that affect the modification of proteins and methods for the screening of protein variants that have different biological activities.

Upstream processing will require research to

• Develop improved devices and procedures for screening the biological activities of new proteins or other compounds. This might include prolonged primary cultures of organisms, tissues, groups of cells, or individual cell lines and in vitro procedures using relevant portions of the immune system to assess the immunogenicity of new or modified compounds.

• Develop technology that improves the stability of proteins and cells that are exposed to dynamic gas-liquid interfaces.

• Develop on-line assay technology that recognizes the quality and quantity of recombinant proteins.

• Engineer containment systems for use in manufacturing.

• Design sterile, low-shear pumping capacity for large-scale recycle reactors.

• Develop predictive models for scaleup of stirred tank and air-lift reactors that consider potential shifts from kinetic to transport limitations on scaleup.

Challenges in *downstream processing* are to

• Develop high-resolution protein purification technologies that offer the degree of resolution provided by reverse-phase HPLC, but that are easily scalable and do not require solvents that are expensive and difficult to dispose of.

• Improve technologies for protein separation based on molecular size. The technologies should provide good resolution and inexpensive operation (i.e., have high specific-throughput rates).

• Develop rapid on-line and off-line assay capabilities for assessing contaminant host proteins and DNA at parts-per-million concentrations, variants of the product protein, and relative biological activity.

• Integrate rapid monitoring, on-line biospecific product detection, and multistep purification sequences to obtain automated purification systems.

• Develop integrated computer-design algorithms for computer-based optimization of purification sequences.

Other technologies that need to be developed include methods for

• Stable liquid formulation of protein pharmaceuticals.

• Precise modifications of proteins to produce homogeneously modified complex proteins with precise biological effects.

• Nonparenteral administration of protein pharmaceuticals, including sustained-release capability; some pharmaceuticals might also benefit from deliverability in a pulsatile manner.

• Characterization of operational and capital costs to allow accurate and rapid estimation of overall process cost.

Most biological manufacturing (at least of high-value products) is now carried out in small lots. As biopharmaceuticals grow in volume and scale and as cost becomes more important in long-term profitability, the development of efficient, integrated manufacturing systems will be more important in bioprocess engineering. All the issues now discussed in other types of manufacturing—design for manufacturing, intrinsic manufacturability, integrated systems of sensors and controls, and integrated information-processing systems for manufacturing—will become relevant for bioprocessing. Development of the appropriate, sensored systems is probably the key element in the early stages of efficient computer-managed biological manufacturing systems.

4.2 SPECIALTY BIOPRODUCTS AND INDUSTRIAL CHEMICALS

Specialty products are chemicals, proteins, microbial substances, and other biologically derived materials whose volume of annual domestic or worldwide use is measured in tons. Specialty products tend to have sale prices less than 3 times their manufacturing costs. Human therapeutics, in comparison, have sale prices 10-12 times their manufacturing costs and annual volumes of use measured in kilograms to hundreds of kilograms.

Specialty chemicals and biologicals are further defined to include antibiotics; food products, additives, and processing aids; oxygenated chemicals and fuel additives; biological agents used in agricultural and environmental applications; value-added products derived from agricultural commodities and other renewable resources; and energy-related products (OTA, 1991). Biotechnology will affect the formulation of products in agriculture, energy, the environment, and human health and has potential to surpass the computer industry in size and importance, because of the pervasive role of biologically produced substances in everyday life (Council on Competitiveness, 1991). Bioprocess engineering is an essential component for rapid transition of bioproducts from the laboratory to a manufacturing scale able to provide the benefits of biotechnology on a large scale at a reasonable cost.

The opportunities for biotechnology products to affect the U.S. chemical industry are substantial. In 1990, chemical shipments were estimated at about $300 billion, and the chemical industry employed a million people and produced more than 50,000 chemicals and formulations (OTA, 1991). According to the Office of Technology Assessment (OTA), biotechnology will likely be used in the chemical industry in the production of fermentation-derived chemicals and synthesis of complex chemicals. It is envisioned that improvement of production processes used by major chemical companies will "be introduced without the fanfare that has accompanied other biotechnology developments." The impact of biotechnology is expected to be incremental and unheralded and result in improvements in productivity when, for example, enzymes are used to replace difficult or expensive steps in chemical synthesis (OTA, 1991).

4.2.1 Enzyme Technology and Specialty Bioproducts

Specialty products derived from biotechnology processing range from low-value commodity materials, such as fuel ethanol, to catalytic enzymes derived from recombinant *Bacillus* species for use in detergents. The worldwide enzyme market is estimated to be about $650 million, of which 50% is attributed to enzymes used in making detergents (Layman, 1992). That group of products is sensitive to processing costs, and manufacture is carried out at a scale in which bioprocess engineering is an important component of technology and design. Specialty products also include flavor enhancers, such as monosodium glutamate, and amino acids. In some cases, recombinant technology has been used to improve the productivity of microorganisms in what is otherwise a mature art of the fermentation industry.

4.2.2 Biopesticides

The search for biodegradable and environmentally compatible pesticides will affect the markets dominated by synthetic insecticides, of which $2 billion worth is sold annually in the United States. Recent examples are *Bacillus thuringiensis* (BT) insect toxins (produced by Pfizer for Ecogen's biopesticide products) and azadirachtin derived from the oil of neen tree seeds (Anonymous, 1992; Stone, 1992). Alternative strategies for BT toxins are to introduce the genes for the toxins (proteins) into other microorganisms that are found in parts of plants attacked by pests and to introduce an insect-resistance gene directly into a plant (OTA, 1991; Crawford, 1988). Sales of BT products have grown from $2.4 million in 1980 to $10.7 million in 1989 and are expected to grow at 11%/year (Feitelson et al., 1992). Another important biopesticide will result from the use of baculovirus; this technology has not been realized, but is clearly imminent (see Chapter 6).

4.2.3 Microalgae and New Chemicals

Although over 50,000 microalgal species are known, fewer than 10 have been studied intensively or commercially exploited (Behrens and Delente, 1991). Recent efforts to cultivate phototropic microalgae in bench- and pilot-scale photobioreactors have yielded an exciting new group of organisms to screen for novel chemical entities. The National Cancer Institute has awarded screening contracts for cyanobacteria and protozoal microalgae; more than 5% of microalgal strains have shown initial activity in primary screens. Microalgae hold promise as an untapped source of new chemicals and will certainly be used in future manufacturing processes.

4.2.4   Plant-Cell Tissue Culture

Plant-cell tissue culture is used commercially in Japan to produce the pigment shikonin and ginseng as a health food. In Germany, a 75,000-liter facility has been built and used to produce an immunoactive polysaccharide, although it is not a commercial process. In Israel, bioreactors form an integral part of a micropropagation process used for commercial production of 14 plant species. In the United States, there are no commercial plant-cell culture processes (Payne et al., 1991). However, the production of taxol, a chemotherapeutic, by plant-cell culture is being actively pursued by at least two companies. Taxol is in clinical trials against a wide range of cancers. Taxol is critically scarce, because the current source is the bark of the Pacific yew tree–an uncommon and very slow-growing tree. Alternative sources are essential, and plant-cell culture is one of the potential solutions to the supply problem.

4.2.5   Research Needs and Opportunities

The examples of current manufacturing technology for bioproducts just cited illustrate the wide array of processing steps that are involved in their manufacture. The economics of the individual steps are reflected in product quality, cost, and final application. A wide array of biosciences are required to develop transformed microorganisms, new biocatalysts, and a fundamental understanding of interactions of proteins with their environments; these all go into the final synthesis of ideas and techniques for production of new molecules or for new methods for producing existing bioproducts. The diversity of bioprocess-engineering skills that must be applied to such a wide array of bioproducts is often overlooked. Challenges that bioprocess engineering faces include specialty-equipment design that meets regulatory, biological, and economic constraints; integration of manufacturing processes into environmentally acceptable and economically feasible process concepts; and rapid purification and monitoring of purification processes to obtain high quality, high purity, and consistent output.

Many specialty products are obtained through fermentation in dilute solution. There is an important opportunity to improve the economics of their production by improving the energy efficiency and selectivity of removal of water. Energy-efficient methods for recovering these products from dilute aqueous solutions are needed to reduce the cost of their synthesis and handling and the volume of downstream processing.

Improvements in the design of bioreactors and in conditions for cultivating microorganisms and obtaining products are also needed. Both solid-substrate and submerged fermentations present major opportunities for biomanufacturing and improvement through bioprocess engineering. In the case of solid-substrate fermentations, the application of enzymes such as would be used in biopulping requires efficient solid-liquid contact. Bioprocess-engineering solutions are needed to enable the penetration of enzyme solutions or microbial inoculations into solid material, which must be rapidly processed on a relatively large scale. For submerged fermentations, product inhibition presents an important challenge to bioprocess engineers, because it limits both the rate and the extent of product formation. Novel bioreactor designs, membrane-separation technologies, and processing aids could all improve the economics of producing specialty bioproducts. Another approach is through cell engineering to alter the fundamental physiology of microbial membranes and to moderate the effects of otherwise toxic extracellular fermentation products.

The challenges and opportunities of bioprocessing technology for the manufacture of specialty bioproducts are similar to those related to biopharmaceuticals, except for the special hazards of pathogens and the specialty-product emphasis on economics of production–the lower the value of the product, the more intense the economic focus.

4.3 ENVIRONMENTAL APPLICATIONS

The application of biotechnology on an environmental scale encompasses subjects ranging from pollution control and bioremediation to mining. Inexpensive yet effective methods are being sought for cleaning up hazardous wastes and other contaminants. Present technologies used are landfilling and incineration, but these are becoming increasingly unpopular, expensive, and difficult to institute, because of stricter regulations and public objections. The magnitude of the problem is seen in the amount of national expenditures for environmental cleanup, estimated to be $115 billion for 1990 (Carlin, 1990). Forecasts for the future are even higher. There is a need for more effective technology that uses such organisms as bacteria or such constituents as enzymes (Hinchee and Olfenbuttel, 1991a,b). Biological systems have been used to degrade or transform objectionable chemicals and materials into more environmentally benign substances for years on a very large scale. For example, today's municipal wastewater-treatment plants use bioprocess-engineering principles to dispose of sewage and to provide clean and safe drinking water. Composting is a practice known to many, including weekend gardeners, and is a use of microorganisms to degrade gardening and other wastes. The recent attention to the environment has focused some bioprocessing technology on the transformation of hazardous wastes and the use of biological processes that produce desired products but little or no waste byproduct.

Current bioproducts with markets of environmental importance can be grouped into reagents for pollution control, agriculture, mining, and oil recovery. An excellent summary is found in the recent OTA report *Biotechnology in a Global Economy* (OTA, 1991), and we quote from that document later.

4.3.1 Bioremediation

Bioremediation refers to the use of entire organisms (mostly soil microorganisms) or selected constituents of microbial cells (mostly enzymes) for chemical transformations. Bioremediation transforms a toxic substance into a harmless or less toxic substance. Ideally, the toxic substance is transformed into carbon dioxide and water. If the toxic substance contains a metal or a halogen, such as chlorine or fluorine, there will be additional side-products (perhaps the free metal atom or its ion or a halide ion). *Mineralization* is the term used to describe the complete degradation of a chemical substance to water and carbon dioxide. *Bioaugmentation*, another frequently used term, involves the deliberate addition of microorganisms that have been cultured, adapted, and enhanced for specific contaminants and conditions at the site. Microorganisms used in bioremediation include aerobic (which use free oxygen) and anaerobic (which live only in the absence of free oxygen). Aerobic microbes have been the organisms of choice for degrading hazardous wastes.

Bioremediation is practiced in two modes—in situ and ex situ. In situ bioremediation involves the use of microorganisms to degrade wastes at the site (both on and below the surface) and avoid excavation of contaminated soil and transfer to different locations. Surface remediation is used to treat the top parts of the soil through aeration by the addition of microorganisms, nutrients, and water. Subsurface bioremediation uses microorganisms already in the soil and groundwater and adds oxygen and nutrients. Ex situ treatment involves the excavation of contaminated soil and its transfer to appropriate treatment sites, i.e., bioreactors. The contaminated soil is aerated and treated with nutrients to provide an active environment for the microorganisms of choice. Treatment continues until the soil is sufficiently clean and can be returned to the site. Ex situ techniques are varied but can involve slurry-phase treatments that combine contaminated soil or sludge in bioreactors or solid-phase treatments that involve placing contaminated soils in lined treatment beds. Bioremediation of water or leachate includes treatment with special bioreactors or filters that contain an active film of microorganisms. The choice of method involves many factors, including the contaminant, the site, and the costs that can be borne. Ex situ treatment is usually very expensive (e.g., $100-1,000 per cubic meter of soil).

Most often, the microorganisms are expected to reproduce in situ. Encouraging in situ reproduction is a challenge that is being addressed in small experiments. The only major attempts have been for oil spills, particularly the combined Environmental Protection Agency-Exxon tests with fertilizers in Prince William Sound to clean beaches of the Valdez oil spill. The results were encouraging, although not spectacular (OTA, 1991, p. 134). The OTA report concludes (p. 140):

> Although bioremediation offers several advantages over conventional waste treatment technologies, several factors hinder widespread use of biotechnology for waste cleanup. Relatively little is known about the scientific effects of micro-organisms in various ecosystems. Research data are not disseminated as well as with research affecting other industrial sectors. This is caused by limited Federal funding of basic research and the proprietary nature of the business relationships under which bioremediation is usually used. Regulations provide a market for bioremediation by dictating what must be cleaned up, how clean it must be, and which cleanup methods may be used; but regulations also hinder commercial development due to their sheer volume and the lack of standards for biological waste treatment.
>
> Although some research is being conducted on the use of genetically engineered organisms for use in bioremediation, today's bioremediation sector relies on naturally occurring micro-organisms. Scientific, economic, regulatory, and public perception limitations that were viewed as barriers to the development of bioremediation a decade ago still exist. Thus, the commercial use of bioengineered micro-organisms for environmental cleanup is not likely in the near future.

The report also summarizes the current prospects for genetically engineered micro-organisms in bioremediation (OTA, 1991, p. 139):

> Some basic research is underway on the use of genetically engineered microbes for waste cleanup. The first out-of-laboratory applications of genetically engineered microbes for waste cleanup will be done in bioreactors, because conditions for microbial survival and monitoring are easier to control in a closed system than in an open field. Today's bioremediation sector continues to rely on naturally occurring micro-organisms. Due to scientific, economic, regulatory, and public perception reasons, the imminent use of bioengineered micro-organisms for environmental cleanup is not likely to happen in the near future. More needs to be learned about naturally occurring microbes—much less those that are genetically engineered. The lack of a strong research infrastructure, the predominance of small companies, the lack of data sharing, and the existence of regulatory hurdles all serve as dominant barriers to commercial use of genetically engineered organisms.

Bioremediation promises lower costs than other types of technology for cleaning up the environment. No academic or regulatory agency has published an analysis of the costs of biological treatment compared with other technologies, such as incineration. The only available information is in individual companies' marketing materials. The present committee recommends analysis of that type.

4.3.2 Point-of-Source Biocontrol

A worldwide industry of pollution biocontrol in sewage treatment has existed since the nineteenth century. The various clean-water acts in the United States have stimulated technology advances in the industry since World War II. We will not cover that special application further here, except to the extent that modern biotechnology and transfer of new bioprocessing technology might affect it. A perhaps-attractive application of biotechnology and bioprocess engineering is in point-of-origin control of pollutants before they disperse into the environment.

Many industries–such as the medical industry, electronics, and polymers– are important sources of waste solvents. Their hazardous disposal has caused much groundwater pollution and is a major market for in situ bioremediation. An appealing concept is to offer those sources a biocontrol process at their own sites that could eliminate the environmental hazard. No such processes are on the market, except the traditional sewage-treatment facility. Because of the characteristic biochemical variability of the influx, such systems often fail. Given the importance of the subject to both the environment and the industries, the committee recommends a study of it.

4.3.3 Agriculture

Potential environmental applications of genetically engineered organisms in agriculture are varied (see Table 4.5). Genes have been introduced into several plant species to confer resistance to or tolerance of particular herbicides. Plants have also been better engineered to resist disease and pests. Most DNA work on animals focuses on altering livestock, poultry, or fish to improve reproductive performance, weight gain, or disease resistance.

Planned introduction of genetically engineered organisms into the environment, often called *deliberate release*, was the focus of an earlier OTA report.

4.3.4 Mining

Natural microorganisms have been used for mineral leaching and metal concentration. No federal funding directly supports microbiological mining, however, and commercial activity is sparse.

Some international research in biohydrometallurgy is proceeding in Canada, South Africa, the United Kingdom, and the United States. The Canadi-

Table 4.5 Some Potential Uses of Biotechnology in Agriculture

(a) Microorganisms

Bacteria as pesticides:

"Ice-minus" bacteria to reduce frost damage to agricultural crops
Bacteria carrying *Bacillus thuringiensis* toxin to reduce loss of crops to dozens of insects
Mycorrhizal fungi to increase plant growth rates by improving efficiency of root uptake of
    nutrients
Nitrogen-fixing bacteria to increase nitrogen available to plants and decrease the need for
    fertilizers

Viruses as pesticides:

Insect viruses with narrowed host specificity or increased virulence for use against specific
    agricultural insect pests, including cabbage looper, pine beauty moth, cutworms, and
    other pests

Vaccines against animal diseases:

Swine pseudorabies
Swine rotavirus
Vesicular stomatitis (cattle)
Foot and mouth disease (cattle)
Bovine rotavirus
Rabies
Sheep foot rot
Infectious bronchitis virus (chickens)
Avian erythroblastosis
Sindbis virus (sheep, cattle, chickens)

(b) Plants

Herbicide resistance or tolerance to:

Glycophosphate
Atrazine
Imidazolinone
Bromoxynil
Phosphinotricin

Disease resistance to:

Crown gall disease (tobacco)
Tobacco mosaic virus

Pest resistance:

BT-toxin protected crops, including tobacco (principally as research tool) and tomato
Seeds with enhanced antifeedant content to reduce losses to insects while in storage

Enhanced tolerance to environmental factors, including:

Salt
Drought
Temperature
Heavy metals

continued

Table 4.5   *Continued*

Enhanced marine algae:

Algae enhanced to increase production of such compounds as B-carotene and agar or to
    enhance ability to sequester heavy metals (e.g., gold and cobalt) from seawater

Forestry:

Trees engineered to be resistant to disease or herbicides, to grow faster, or to be more
    tolerant to environmental stresses

(c) Animals
Livestock and poultry:

Livestock species engineered to enhance weight gain or growth rates, reproductive
    performance, disease resistance, or coat characteristics
Livestock animals engineered to function as producers for pharmaceutical drugs

Fish:

Triploid salmon produced by heat shock for use as game fish in lakes and streams
Fish with enhanced growth rates, cold tolerance, or disease resistance for use in aquaculture
Triploid grass carp for use as aquatic weed control agents

SOURCE: OTA, 1991.

an Center for Mineral and Energy Technology is the leading government
research agency in the field. One focus for the Canadians is uranium bioleaching; one mine is now bioleaching 90,000 lb of uranium per month.
The biological mitigation of acidic mine drainage is another Canadian
project. Research is slow, however, because of the economics of the mineral market. As long as metals are plentiful and easily mined, no economic
advantage is realized by microbiological mining.

4.3.5 Microbial-Enhanced Oil Recovery*

It has been estimated that more than 300 billion barrels of U.S. oil
cannot be recovered by conventional technology and might be accessible
through enhanced oil production. That volume is 2.5 times as large as the
amount of oil produced by the United States since 1983. The actual enhanced oil recovery has been low—no greater than 5% of total U.S. production, even though various Department of Energy incentives have been available. Other countries, such as Canada, have projected that by the year 2010
one-third of its oil recovery will use enhanced techniques. In recent years,

*Adapted from the OTA (1991) report.

advanced oil-drilling techniques have enhanced overall yield, and it is expected that these techniques, not microorganisms, will satisfy oil companies' needs for greater yield in the short term.

Although most of the major oil companies have in-house staff investigating and perfecting microbial-enhanced oil recovery (MEOR), the methods' low cost might appeal more to small-field operators, who have already pumped and sold the easy-to-get component of their fields. MEOR is not predictable; like the use of microorganisms for hazardous-waste remediation, the use of microorganisms for oil recovery is site-specific. Individual oil deposits have unique characteristics that affect the ability of microorganisms to mobilize and displace oil. An understanding of the microbial ecology of petroleum reservoirs is a prerequisite to the development of any MEOR process, whether microbial or not, because an inappropriate design might accelerate the detrimental activities of microorganisms (e.g., corrosion, reservoir souring, and microbial degradation of crude oil). Basic environmental-biotechnology research under way for contaminated soil and groundwater will provide much needed information to those working on MEOR, who face several serious challenges (OTA, 1991):

• Better biochemical and physiological understanding of microorganisms already present in oil reservoirs.
• Development of microorganisms that degrade only less-useful components of oil.
• Screening of microorganisms for production of surfactants and viscosity enhancers and decreasers.

4.3.6   Research Needs and Opportunities

The committee concurs with the OTA report (OTA, 1991) that immediate opportunities for bioprocessing, particularly those which would use genetically engineered microorganisms, exist. The impact of bioprocessing on environmental remediation and industrial waste control could be tremendous over the longer term. The technical aspects of environmental issues are broad and complex and the technical elements of the opportunities ill-defined, and the committee recommends that a study be carried out to set priorities.

4.4 REFERENCES

Anonymous. 1992. Ecogen, Pfizer sign production agreement. Chem. Eng. News 70(3):7.
Bailey, J. E. 1991. Toward a science of metabolic engineering. Science 252:1668-1681.
Behrens, P. W., and J. J. Delente. 1991. Microalgae in the pharmaceutical industry. Bio-
    Pharm 4(6):54-58.

Buchner, J., and R. Rudolph. 1991. Renaturation, purification and characterization of recombinant F_{ab}-fragments produced in *Escherichia coli*. Bio/Technology 9:157-162.

Carlin, A. 1990. Environmental Investments: The Cost of a Clean Environment. EPA-230/ 12-90/084. Washington, D.C.: U.S. Environmental Protection Agency.

Council on Competitiveness. 1991. Report on National Biotechnology Policy, Washington, D.C.

Crawford, M. 1988. ARS prodded into the open. Science 239:719.

Davis, G. T., W. D. Bedzyk, E. W. Voss, and T. W. Jacobs. 1991. Single chain antibody (SCA) encoding genes: One-step construction and expression in eukaryotic cells. Bio/ Technology 9:165-169.

Erlich, H. A., D. Gelfand, and J. J. Sninsky. 1991. Recent advances in the polymerase chain reaction. Science 252:1643-1650.

Feitelson, J. S., J. Payne, and L. Kim. 1992. *Bacillus thuringiensis*: Insects and beyond. Bio/ Technology 10:271-275.

Goloubinoff, P., A. Gatenby, and G. H. Lorimer. 1989. GroE heat-shock proteins promote assembly of foreign prokaryotic ribulose biphosphate carboxylase oligomers in *Escherichia coli*. Nature 337:44-47.

Hammond, P. M., T. Atkinson, R. F. Sherwood, and M. D. Scawen. 1991. Manufacturing new-generation proteins: Part 1. The technology. BioPharm 4(4):16-22.

Hinchee, R. E., and R. F. Olfenbuttel, eds. 1991a. In Situ Bioreclamation: Applications and Investigations for Hydrocarbon and Contaminated Site Remediation. Boston: Butterworth-Heinemann. 648 pp.

Hinchee, R. E., and R. F. Olfenbuttel. 1991b. On-Site Bioreclamation: Processes for Xenobiotic and Hydrocarbon Treatment. Boston: Butterworth-Heinemann. 560 pp.

Isogai, T., M. Fukagawa, I. Aramori, M. Swami, H. Kojo, T. Ono, Y. Ueda, M. Komsaka, and H. Imanaka. 1991. Construction of a 7-aminocephalosporanic acid (7ACA) biosynthetic operon and direct production of 7ACA in *Acremonium chrysogenum*. Bio/Technology 9:188-191.

Layman, P. 1992. Promising new markets emerging for commercial enzymes. Chem. Eng. News 68(39):17-18.

Marino, M. H. 1989. Expression systems for heterologous protein production. BioPharm 2(7):18-20, 22, 24-26, 28-29, 32-33.

Mitraki, A., and J. King. 1989. Protein folding intermediates and inclusion body formation. Bio/Technology 7:690-697.

OTA (Office of Technology Assessment). 1991. Biotechnology in a Global Economy, B. Brown, ed. Office of Technology Assessment, U.S. Congress, Report No. OTA-BA-494. Washington, D.C.: U.S. Government Printing Office.

Payne, G. F., V. Bringi, C. L. Prince, and M. L. Shuler. 1991. Plant Cell and Tissue Culture in Liquid Systems. New York: Hanser Publishing.

Pelham, H.R.B. 1986. Speculations on the functions of the major heat shock and glucose-regulated proteins. Cell 46:959-961.

Ruettimann, K. W., and M. R. Ladisch. 1987. Casein micelles: Structure, properties and enzymatic coagulation. Enzyme Microb. Technol. 9:578-589.

Stone, R. 1992. A biopesticidal tree begins to bloom. Science 255:1070-1071.

Waldmann, T. A. 1991. Monoclonal antibodies in diagnosis and therapy. Science 252:1657-1662.

5

Needs: What Must Be Done to Meet the Challenges

5.1 EDUCATION AND TRAINING

Research activities and the training of bioprocess engineers for the next decade should be broad enough to enable staffing of bioprocess research, development, and manufacturing functions for biotherapeutics and other classes of bioproducts, including intermediate-value products obtained from renewable resources through bioprocessing, value-added agricultural materials, and waste-processing products and services. This chapter treats elements of bioprocess engineering that must be addressed to meet the needs of industry and the goal of commercializing biotechnology.

5.1.1 Science–Engineering Interface

The principles, culture, and techniques of scientists (biologists and chemists) are often different from those of bioprocess engineers. The differences can place unnecessary limits on collaboration among members of a bioprocessing-development team and thereby delay engineering considerations to the later stages of bioprocess development. Hence, it is important that the bioprocess engineers' training in the next decade have a strong background in biochemistry, molecular biology, cell biology, and genetics. That will facilitate useful communication of bioprocess engineers with the bench scientists who are at the initial discovery stage of biological-product research and development. The situation can be thought of as analogous to process development in the chemical and petrochemical industries, where engineers who are knowledgeable in basic concepts in organic and physical chemistry have fostered innovations in processes developed through interactions of research chemists and engineers. Future research and training in the appli-

cation of systems engineering and bioprocess economics at the early stages
of research and development are also needed; they will help to foster interactions between bioscientists and bioprocess engineers and encourage con-
sideration of engineering factors at the onset of a research and development
program.

5.1.2 Multidisciplinary Team Research

Bioprocess engineering is a broad engineering field, in that it covers all
the physical sciences and biological sciences. It is impossible to design and
engineer bioprocesses within a single discipline. Formal coursework in
other disciplines will begin to build the foundation for team research through
cross-disciplinary interactions. However, coursework alone will not be sufficient for executing and implementing the actual research and development. The hands-on experience of team research must be part of bioprocess
engineers' training program.

It is therefore recommended that cross-disciplinary research be part of
the training of the bioprocess engineer; it is probably best practiced at the
postgraduate level. There is much to be gained through input from different
disciplines when such team research is executed. However, to implement
this type of research, cultural changes in the engineering and scientific
communities will be required. For example, a doctoral candidate in chemical engineering is often viewed as performing research as a single investigator when, in fact, input from multiple disciplines is essential.

Several government-agency programs foster cross-disciplinary and interdisciplinary training. They include the National Science Foundation (NSF)
Engineering Research Center Initiative and the National Institutes of Health
(NIH) Interdisciplinary Biotechnology Training Grant Program in the National Institute of General Medical Sciences. It is recommended that the
programs continue to foster activities through cross-disciplinary interactions.

5.1.3 Industry–University Interface

The education of leaders who are strong in science, engineering, business, and management skills is difficult. But the bioprocess engineer's
education at the predoctoral level usually devotes little time to the management and business aspects of biotechnology. A training program must re-
flect the realities of the bioprocess industrial sectors. It is recommended
that future programs incorporate the industry-university interface into formal training activities.

Continuing education is especially critical for bioprocess engineering,
because of the rapidity of advances in the biological sciences. Continuing

education in the industrial sector should be part of the training offered by universities to leaders in the bioproduct industry. Such a program should be created by industry, universities, and government in a cooperative fashion.

5.1.4 Bioprocess-Equipment Engineers

Biochemical engineers are and should be the lead engineers in bioprocess development. They are educated uniquely to span the gap between the biological sciences and process engineering. However, the efforts of these engineers must also be integrated with those of equipment engineers, as well as bioscientists.

Integration of biochemical and equipment engineering is often absent in current bioprocess-engineering practice. The equipment that is used by the bioprocess engineer evolves slowly; there are few radical breakthroughs. That is probably because the most active parts of the industry are new and relatively small, and their progress has been driven by new ventures attempting to inject new technology into bioprocessing. Attempts by small, startup bioprocess-equipment companies are often underfunded, and many fail. In contrast, the well-established manufacturers of bioprocess equipment spend little on research and development relative to other high-technology industries.

Systems and equipment engineering, comparable with that in the aircraft, electronics, and defense industries, must be used by the U.S. bioproducts industry to make it competitive. To that end, more engineers should be trained who are seriously interested in improving bioprocess equipment, such as chromatography systems, centrifuges, membrane filters, bioreactors, and especially on-line instrumentation for monitoring and control. Their undergraduate education can be in the traditional fields of instrumentation and electrical and mechanical engineering, with a few basic courses in chemical engineering. At the graduate level, education should be structured jointly with programs in bioprocess engineering.

As the biotechnology industry matures and manufacturing costs become important, the equipment engineer will assume a larger role than today. And, if biomass substantially replaces fossil fuel as a primary source of energy and materials, equipment technology will become critical. Industry and government should encourage the education of more equipment engineers for the bioproducts industry.

5.1.5 Diversification and Specialized Training

Bioprocess-engineering manpower demand will continue to increase in research and development, manufacturing, biotechnology-related business and legal professions, and teaching. Specialized training with a focus on specialties will be required. Specialties in the biopharmaceutical arena are

- Recombinant and nonrecombinant fermentation technology.
- Bioseparation and purification of gene products.
- Animal-, plant-, and insect-cell culture and animal-tissue culture.

Similarly, bioprocessing of renewable resources and environmental bioprocess engineering requires

- Recombinant and nonrecombinant fermentation technology.
- Bioseparation and purification of fermentation products or products derived through microbial activity.
- Applied microbiology of industrially important microorganisms.

Other fields of importance are

- Protein chemistry and processing.
- Biocatalysis and enzyme technology.
- Biosensors, instrumentation, and process control.

Future training and educational programs must be much more concentrated and focus on selected subspecialties to address anticipated staffing needs in the pharmaceutical industry, medical industry, food and agriculture, environmental biotechnology-related industry, chemical industry, and energy industry.

5.1.6 Curriculum Development

A unique element in the education of some bioprocess engineers is hands-on experience in applied microbiology and molecular biology, bioreactor operation, cell culture, bioseparation (chromatography, membranes, and centrifugation), and basic analytical methods for biological materials and molecules. Such training in the context of an upper-level, undergraduate bioprocess-engineering laboratory constitutes an invaluable first experience in merging theory with experiment for biological systems. It is already being provided to some extent in a few universities.

The committee recommends that competitive-grant programs be further developed to upgrade teaching laboratories for bioprocess engineering so that they can provide a high-quality training experience for a larger number of students.

5.2 RESEARCH

Bioprocessing-research needs differ between the biopharmaceutical, renewable-resources, and environmental sectors of the biotechnology industry. The biopharmaceutical sector currently has the strongest basis and research needs directly tied to recombinant-DNA technology. Key research needs in generic applied research are in fundamental studies on and development of

• Methods for rapid characterization of biochemical properties, efficacy, and immunogenicity of protein pharmaceuticals.
• Process control of systems involving genetically altered products.
• High-resolution protein-purification technologies that are economically feasible, are readily scaled up, and have minimal waste-disposal requirements.

Other research needs are related to expanding the range of pharmaceuticals that can be produced by prokaryotic cells, further developing the technology for stable liquid formulations and for sustained release of protein pharmaceuticals, and increasing knowledge of chemical and biochemical reactions that modify proteins during production and storage. The latter subjects are perceived to be potentially parts of the research mission of NIH, because they involve health issues. The key bioprocess-engineering issues are consistent with the continuing programs of NSF, although sustained increases in resources will be needed to fund strong programs.

The processing of renewable resources and manufacture of value-added products from agricultural commodities require bioprocess-engineering research to address fundamental understanding and development of

• Cellulose pretreatment and saccharification systems to convert ligno-cellulosic materials, as well as coproducts of corn processing, into appropriately priced fermentable sugars and value-added materials.
• Microorganisms and fermentations capable of converting pentoses to value-added products at rates and yields comparable with those obtained for glucose by yeast.
• Separation systems, amenable to large-scale use, for recovering and purifying bioproducts from dilute aqueous solutions.
• Engineering and manipulating cellular pathways for enhanced production of microbial metabolites, or microbial synthesis, of new products.

Those research needs can be addressed within the framework of bioprocessing research initiatives of NSF, the U.S. Department of Agriculture (USDA), and the Department of Energy (DOE). Another engineering-research need is in the development of large-scale surface culture as might be encountered in biopulping. Other fundamental research needs are to increase the knowledge base of biochemical and microbial transformations that result in value-added nonfood products from starch and cellulose. A recent example is the genetic engineering of *E. coli* to enable it to produce ethanol, thereby combining the microorganism's ability to use pentose with its ability to yield an economically useful end product (Ingram, 1992). Other examples of potential products are poly-β-hydroxybutyrate, calcium magnesium acetate (from acetic acid), glycerol, acetone, and butanol (Bungay, 1992). Research to improve methods of adding value to corn wet-milling products is also needed. Those subjects all fit within the missions of USDA initiatives to carry out applied generic research in biotechnology to add value to agricultural products and DOE programs in deriving fuels and chemicals from biomass. A sustained research effort will be required if successful process concepts are to be developed in the coming 5-10 years.

Environmental applications of bioprocessing are perhaps the furthest off and need fundamental research, particularly in

- Specific effects of microorganisms in various ecosystems.
- The role of microorganisms in ex situ waste remediation.
- Definition and implementation of engineering standards by which bioremediation protocols and processes could be gauged.

The research needs in bioremediation are subject to a number of complex technical and regulatory issues, as described in the OTA report (OTA, 1991). Concerted efforts will be particularly important as the regulatory environment for biotechnology products improves and the regulatory process is streamlined.

The federal support of fundamental research in bioprocess engineering is essential, and a major increase in federal support is strongly recommended. The research goals and approaches of industry are different from those of universities. Industrial research is mission-oriented, and its emphasis is on applied research that leads to products and more efficient process technologies; the purpose of university research is to enlarge the generic and fundamental knowledge base relevant to bioprocess engineering. Consequently, federal support is a critical element of success.

The challenge of coordinating government-supported research and development is recognized and is the focus of the Federal Coordinating Council for Science, Engineering, and Technology (FCCSET) Committee on Life Sciences and Health. The FCCSET report (1992) lists planned federal investments in manufacturing and bioprocessing biotechnology, which total $123.8 million for FY 1993 (Table 5.1).

The committee agrees with the directions that are set, but feels strongly that more will be needed over the next 10 years. Bioprocess technology is the basis on which the products of life-science research are translated to a manufacturing environment. It is critical for bringing the industry to the profitability that will return taxes and create jobs in all sectors of our economy.

Table 5.1 Federal Investment in Manufacturing and Bioprocessing Biotechnology

Agency	Investment, millions of dollars		
	FY 1991	FY 1992	FY 1993
National Science Foundation	29.5	32.3	43.0
U.S. Department of Agriculture	17.6	18.8	23.6
Department of Defense	12.7	17.4	18.4
Department of Health and Human Services	16.7	17.0	17.7
Department of Energy	5.3	6.4	13.2
National Aeronautics and Space Administration	2.6	2.6	3.7
Department of Commerce	3.5	3.5	3.5
Department of the Interior	1.0	0.8	0.7
TOTAL	88.9	98.8	123.8

SOURCE: FCCSET, 1992, p. 46.

5.3 TECHNOLOGY TRANSFER

Technology transfer in biotechnology and bioprocess engineering can include dissemination of published scientific and technical literature related to biotechnology, movement of scientists and engineers between employers, training of scientists and engineers in bioprocessing technology, construction of plants to manufacture biotechnology products, joint ventures of biotechnology businesses, licensing of biotechnology products and bioprocesses, exchange of manufacturing technology, release of technical information with the sale of bioprocessing equipment, technical consulting, transfer of engineering proposals, and transfer of technical information through trade exhibits.

Although there is a need for university, industry, and government research organizations to be interdependent in their research and development endeavors, effective communication and technology transfer are mutually beneficial and critically important to the national economic security. The federal government is emphasizing the need to increase cooperative activities between national laboratories, industry, and universities with emphasis on technology transfer. The committee encourages continued development of this type of interaction.

The important issues that are relevant to biotechnology transfer and should be focused on are the effectiveness of U.S. university-industry relationships in bioprocess-technology transfer in the context of international exploitation of biotechnology.

5.3.1   University-Industry Relationships

The United States seems to have a very effective university-industry technology-transfer process, compared with other technologically advanced countries. The effectiveness of technology transfer from the U.S. university to industry can be attributed to the high degree of freedom of faculty to conduct their research and the openness of the academic community to technology transfer. Forms of technology transfer from the universities include employment of graduates, professional meetings, dissemination of scientific and technical publications, consulting arrangements, contract research for industry, collaborative research agreements, and training of industrial personnel.

Thus, universities make their research available to industry by many means. It appears that, in biological science, industry carefully studies the university output and exploits it effectively. In contrast, the U.S. bioprocessing industry is slow to use information for innovation in manufacturing processes. Some of the industries involved are very conservative about changing processes and, in general, U.S. industry tends to be risk-averse because owners exert pressure to produce results in the short term, and the regulation and validation process could become a rate-limiting factor in the commercialization of pharmaceutical products.

In the bioprocessing industry, many manufacturing-process innovations are developed by small companies. New chromatographic techniques, new bioreactors, new membrane filtration–all come from small to medium companies. The large users of the technology–pharmaceutical, food, and chemical companies–seem to have less of a role in innovations that lead to new manufacturing processes. Research programs that lead to innovations by universities or entrepreneurial companies are supported primarily through government funds. There are no U.S. consortia of manufacturers supporting the development of generic bioprocessing technology.

Both industry and university technology transfer would benefit if industry had stronger communications with the universities. Communication would serve as a means to encourage cooperation between industry and universities to the benefit of both.

The committee recommends that the issue of tax incentives be reexamined with an eye to stimulating greater risk investment by industry, improving technology transfer, stimulating university investigators to set priorities in their fundamental research according to technology that industry will use, and promoting international competitiveness in bioprocess technology.

The committee strongly recommends that a strategy be developed for fostering an improvement in awareness of the importance of manufacturing technology in the research and university communities through education and training. It would be equally beneficial if industry would provide guidance to universities in selecting research and development foci.

5.3.2 International Exploitation

Biotechnology companies have three ways of exploiting biotechnology in a world market: They can license the biotechnology to a foreign company. They can invest in a foreign subsidiary or joint venture. They can make the biotechnology product in their home countries and export it.

Many new biotechnology firms in the United States have transferred some of their biotechnology via licensing and joint ventures to well-established large U.S. or foreign companies, because the new firms lack the capital and capability for commercial-scale manufacturing or marketing abroad. It is difficult to assess accurately the amount or extent of international biotechnology transfer, investment, and trade and their potential impact on U.S. competitiveness in biotechnology. The issue is quite complex and might warrant a separate study.

5.4 REFERENCES

Bungay, H. 1992. Product opportunities for biomass refining. Enzyme Microb. Technol. 14:501-507.

FCCSET (Federal Coordinating Council for Science, Engineering, and Technology). 1992. Biotechnology for the 21st Century, A Report by the FCCSET Committee on Life Sciences and Health, Office of Science and Technology Policy, Executive Office of the President, Washington, D.C.

Ingram, L. 1992. Genetic engineering of novel bacteria for the conversion of plant polysaccharides into ethanol. Pp. 507-509 in Harnessing Biotechnology for the 21st Century, M. Ladisch and A. Bose, eds. Washington, D.C.: American Chemical Society.

OTA (Office of Technology Assessment). 1991. Biotechnology in a Global Economy, B. Brown, ed. Office of Technology Assessment, U.S. Congress, Report No. OTA-BA-494. Washington, D.C.: U.S. Government Printing Office.

6

The Future

Bioprocess engineering is concerned with translating biological science into biologically based manufacturing. To be prepared for the biological manufacturing systems of the future, it is important to identify the fields of science and technology that have reached or will soon reach early prototypes and to begin to develop engineering systems to deal with them. The lead time in development of any new technology is long. Biological science is so prolific that any present list of future developments of technology must be incomplete. It is, however, straightforward to identify subjects in which new engineering techniques must be developed now, if the technologies are to be available when they are needed in large-scale manufacturing 5-15 years from now.

6.1 OPPORTUNITIES

6.1.1 Biopharmaceuticals and Biopesticides from Insect Cell-Baculovirus System

Insect cells from moths can be cultivated in a manner similar to (but not identical with) the manner in which mammalian cells can. Unlike mammalian cells, insect cells are naturally continuous cell lines. They can be adapted easily to growth in serum-free media. The unique biphasic life cycle of the baculovirus, which readily infects cultured insect cells, makes it an ideal vector for expression of foreign genes. The baculovirus contains a late promoter that is very strong—perhaps the strongest eukaryotic promoter known. Consequently, the insect cell-baculovirus expression can be used to produce very high levels of protein (up to 40% of total protein) in a nontransformed host cell (i.e., noncancerous) with a vector that is non-pathogenic to vertebrate animals. Those features of high expression levels and increased safety distinguish this system from the commonly used mammalian expression systems. However, the high expression levels are not typically obtained with secreted, glycosylated proteins. Although insect cells have most of the posttranslational machinery of mammalian cells, proteins produced in the baculovirus system are not processed in precisely the same way as in mammalian hosts (Luckow, 1990; Shuler et al., 1990).

Currently, the insect cell-baculovirus system is widely used to produce research quantities of proteins in many industrial laboratories. It is a convenient system for gene expression, and most proteins are produced in the correctly folded conformation. One product, a coat protein from HIV (the virus responsible for AIDS) produced from the baculovirus system, is in Phase II clinical trials and could be produced commercially within 5 years. Other commercial developments of the baculovirus system will depend on bioprocess research to increase expression levels of secreted glycosylated proteins.

The baculovirus itself has been approved for use as a pesticide. Genetic modifications to the virus to increase its effectiveness (e.g., speed of killing) are being tested. Large-scale production of such a biopesticide with cell culture will present unprecedented challenges for large-scale, inexpensive animal-cell reactor designs.

6.1.2 Gene-Based Pharmaceuticals and Gene Therapy

New classes of products are being tested for use in humans and animals, all sharing genes as common targets. Products based on antisense technology directed toward neutralizing messenger RNA are probably being pursued most vigorously; gene therapy through permanent alteration of chromosomes might hold the greatest potential for treatment of diseases like cancer and for correction of genetic disease. The products depend either on classes of compounds that are related to nucleic acids (oligonucleotides and oligonucleotide analogues), on cells that have been genetically altered, or on viruses that bear appropriate nucleic acids. For the large-scale production of nucleotides and nucleotide analogues, new molecular techniques must be developed. There are now no procedures for making substantial quantities of these types of materials in high purity and with appropriate chirality. Basic chemical and biochemical techniques must be developed for their preparation; new techniques (probably based on high-pressure chromatography) will be required for large-scale purification, and biological methods might be required for preparation of precursors and perhaps for formation of bonds. For genetically modified cells and viruses, the usual techniques for mammalian-cell culture and molecular biology will be required, as will additional measures for safety and for economical, patient-specific production.

Vignette 6

Cell-Transplantation Therapy

Cell transplantation is being explored as a means of replacing tissue function. Individual cells are harvested from a healthy section of donor tissue, isolated, expanded in culture, and implanted in a patient at the desired site of the functioning tissue. Also, cell-based therapies are being developed that involve the return of genetically altered cells to the host with gene-insertion techniques.

Cell transplantation has several advantages over whole-organ transplantation. Because the isolated cell population can be expanded in vitro with cell culture, only a small number of donor cells are needed to prepare an implant. Consequently, the living donor need not sacrifice an entire organ. The need for a permanent synthetic implant is eliminated through the use of natural tissue constituents without the disruption and relocation of a whole piece of normal tissue. The use of isolated cells also allows removal of other cell types that might be the target of immune responses, thus diminishing the rejection process. In addition, major surgery on the recipient and donor, with its inherent risks, is avoided. Finally, the cost of the transplantation procedure can be reduced substantially.

Isolated cells cannot now be made from new tissues in complete isolation. They require specific environments that often include supporting material to act as a template for growth. Three-dimensional scaffolds will probably be used to mimic their natural counterparts, the extracellular matrices (ECMs) of the body. The scaffolds will serve as both a physical support and an adhesive substrate for isolated parenchymal cells during in vitro culture and later implantation.

Because of the multiple functions of the materials, the physical and chemical requirements are numerous. To accommodate a sufficient number of cells for functional replacement, a cell-transplantation device must have a large surface area for cell adhesion. High porosity provides adequate space for cell seeding, growth, and ECM production. A uniformly distributed and interconnected pore structure is important for easy distribution of cells throughout the device and formation of an organized network of tissue constituents. This allows for cell-cell communication through direct contact and through soluble factors. Also, nutrients and waste products must be transported to and from differentiated groups of cells, often in ways that maintain cell polarity. In the reconstruction of structural tissues, such as bone and cartilage, tissue shape is integral to function. Therefore, the scaffolds must be processable into devices of varied thickness and shape. Furthermore, because of the goal of eventual human implantation, the scaffold must be made of biocompatible materials. As the transplanted-cell population grows and the cells function normally, they will begin to secrete their own ECM support. The need for an artificial support will gradually diminish; if the implant is biodegradable, it will be eliminated as its function is replaced.

Two major research thrusts are required to develop technology for cell transplantation. The first deals with appropriate cell-culture techniques, and the second addresses the nature of the scaffold. The two are closely related, and examination of one issue will influence the other. Both require innovative bioprocess engineering to produce differentiated and structured products at a cost that our national health-care system can afford.

The study of the adhesive interactions between cells and both synthetic and biological substrates will be pivotal in determining the effect of different physical and

chemical factors on cell and tissue growth and function. Until recently, most research in the field has focused on minimizing biological fluid and tissue interactions
with biomaterials in an effort to prevent fibrous encapsulation from foreign-body
reaction or clotting in blood that has contact with artificial devices. In short, much
biomaterials research has focused on making the material invisible to the body.
Innovations that use the inverse approach–programmed extensive interaction of the
material with biological tissue–will give biomaterials research a new focus. Novel
biomaterials that incorporate specific peptide sequences will be developed to improve cell adhesion and promote differentiated-cell growth by releasing growth factors, angiogenesis factors, and other bioactive molecules.

Cell-based therapies and artificial organs have the potential to have a great impact in medicine for treatment of diseases of aging, degenerative diseases, burns,
blood and lymphoid disorders, orthopedic problems, and others. A multidisciplinary
approach based on recent advances in biochemistry, cell biology, and materials
science will be necessary to respond to emerging technology problems. Developing
the needed differentiated-cell culture in three dimensions on a large enough scale to
be economically feasible will remain a challenge for bioprocess engineers well into
the twenty-first century.

6.1.3 New Catalysts

New types of catalysts based on biological systems are being developed.
Among them are catalytic antibodies (abzymes) and catalytic nucleic acids
(ribozymes). Those types of materials can, in principle, be used both in
specialty-chemical production and in human therapy. Although the techniques required for preparing abzymes will be the same as those used in
other kinds of monoclonal-antibody manufacture, production of ribozymes
(initially for applications in human health) will require an entire new array
of manufacturing, purification, and production technologies; there are no
large-scale methods for preparation, isolation, and purification of high-mo-
lecular-weight nucleic acids.

6.1.4 Cells, Organs, and Biomaterials

Production of human skin is already a commercial business in the United
States; clonal production of lymphocytes is in an early stage of development. With the enormous advances in the biology of differentiation and
development, a clear target for the future is large-scale production of cells
(initially) and intact organs (later) for use in therapy and organ replacement.
Bioprocess engineering will be required to develop new types of reactors
and to delineate biological mechanisms that affect growth and maintenance
of the cells and differentiated state of these cells. Although lymphocytes in
culture can be prepared now, economical production still requires substantial improvement of existing techniques.

The technology for production of organs will be much more complex. The rapid increase in knowledge concerning the role of growth factors, cytokines, and other molecules important in cellular communication, coupled with the realization that the three-dimensional matrix in which tissue cells grow is crucial in maintaining cell phenotype, has opened up the possibility of cultivation of viable functioning organ structures. The first applications, already well under way, are in bone marrow cultures and tissues for cosmetic reconstruction. With the increasing concern about the safety of blood products, the ability to culture specific types of blood cells might be crucial for treatment of many diseases and wounds and for surgical procedures. There are serious engineering challenges in developing a system for successful large-scale cultures. Important problems remain in understanding the role of mass transfer, protein matrix for cell attachment and growth, and medium formulation, particularly the proper combination of growth factors required to optimize production of specific cell types. Cosmetic applications include cartilage and artificial skin. Great strides in the latter field in the last decade have resulted in many applications, from burn treatment to reconstruction after surgical procedures, such as breast-cancer removal. But there is need for much improvement even here.

Longer-range opportunities include hepatic, pancreatic, and kidney cell cultures. A complex three-dimensional substrate is required to maintain cell differentiation and organ function. Again, many engineering problems must be solved. Providing new vasculature for substrate delivery and product removal is vital. Most organ cells have a polarity that is required for proper function. The overall basic biology of the complex systems required to direct and control differentiating cells is not now understood, and it is impractical to specify in any detail the types of reactors that will be required in the future, other than to say that they will be much more complex and interactive than those now used.

With an aging population, there is increasing interest in biologically compatible materials for replacement of organs, joints, and ligaments and for related applications. Collagen, biologically derived polyesters, hyaluronic acid, and other materials have all shown attractive properties in some applications. New biological materials (for example, spider silk or protein adhesives from barnacle and mussel) might be usable for such applications as sutures and bioadhesives. They are "specialties," rather than true pharmaceuticals, even if they are used in human health-care applications. Thus, the economics of production are more important for them than for conventional drugs, and the process aspects of their production are critical.

6.1.5 Transgenic Animals

Transgenic animals are being developed for a wide variety of applications, and bioprocess engineering will play a role in the use of these animals in several ways. If transgenic animals are used as factories for production of biological substances (e.g., as protein in milk), bioprocess engineering will be required to develop appropriate techniques for isolation and purification of the desired products. More important in the short term is that transgenic animals might provide test beds for proving the safety of new pharmaceutical entities and for accelerating their passage through the regulatory process. Bioprocess engineering can play an important role in controlling the testing technologies, perhaps in maintaining transgenic tissues and organs, and in coupling the testing available for use in transgenic animals with the development of processes that are acceptable and robust from a regulatory viewpoint.

6.1.6 Transgenic Plants

Transgenic plants are capable of generating specialty chemicals or other bioproducts. Special bioprocessing capabilities will then also need to be developed for extracting, concentrating, and purifying such products from plant tissue. This sector of bioprocess engineering might also be important to the prospects of expanding crops or developing new varieties that are rich in fermentable carbohydrates, which are readily used as feedstocks for large-scale manufacturing of specialty and industrial chemicals.

Transgenic tobacco plants have been developed to produce monoclonal antibodies identical in function with the original mouse antibody. Other proteins produced in plants are human serum albumin and enkephalins (Hiatt et al., 1989; Hein et al., 1991). Processes to recover and purify proteins from plant-cell extracts will be needed if such systems are commercialized.

6.1.7 Nontraditional Organisms

Many unusual chemicals or enzymes with unique properties come from organisms that are difficult to culture effectively. In particular, marine microbes (especially algae) and extremophiles (primarily the archaebacteria) present important, but long-range opportunities. Bioprocess engineers have already demonstrated the ability to design devices and protocols to culture microbes from deep-sea vents and undoubtedly have the skills necessary to develop techniques for other difficult-to-culture organisms.

6.1.8 Energy and Renewable Resources

A broad range of technical opportunities might require large-scale engineering. For example, if biologically produced materials (surfactants and viscosifiers) are used in enhanced oil recovery and in transportation of heavy crudes and coal slurries, appropriate biological manufacturing facilities must be developed. Particularly for those purposes, the ability to produce materials economically *on site* from local raw materials might be important and might require development of completely new technology. Although ideas for desulfurization of coal and liquid hydrocarbons seem, at present, to be improbable, the development of materials from biological processes to aid in transportation and burning of coal, for example, seems plausible.

The largest-volume process that bioprocess engineering might be called on to address would be consumption of carbon dioxide (CO_2) from combustion. If it is necessary to reduce CO_2 emissions from stationary sources substantially to ameliorate the greenhouse effect, a plausible approach would be to couple CO_2 release with growth of a CO_2-requiring organism (either photosynthetic or nonphotosynthetic). The scale of any facility that would be used in this type of application would be enormous and would require innovative engineering to minimize costs.

In addition to those large-scale processes, there are plausible uses for biological catalysts on a smaller scale in a number of fields related to energy production and bioprocessing. For example, fuel cells based on enzymatic catalysis would provide an attractive method of using ethanol and perhaps other biologically derived fuels with high thermodynamic efficiency in an environmentally acceptable way. A range of chemicals can, of course, in principle be produced from renewable resources. The process for production of acrylamide from acrylonitrile has re-emphasized the practicality of large-volume production of some types of commodity chemicals with enzymatic catalysis (Nagasawa and Yamada, 1989). If energy costs continue to go up, if petroleum feedstocks become more expensive or more erratic in supply, and if processes based on conventional nonbiological catalytic systems become unacceptable from an environmental point of view, biological processing might be able to overcome what is usually an intrinsic economic disadvantage. In that event, bioprocess engineering will be called on to provide reactors and control systems appropriate for the production of commodity products (chemicals and fuels) with tight environmental and economic constraints.

Bioprocess engineering is an essential component for addressing those challenges. Fundamental understanding of the changes that occur in pretreated cellulose to make it more reactive and identification of inexpensive approaches to the engineering of large-scale processes are first steps in producing a reactive substrate. Improvement of both the rate and the extent of xylose fermentation to ethanol requires an understanding of microbial physiology and of the optimal control of fermentation conditions to maximize productivity and yield. The fractionation of solutes from water is a challenge common to most fermentations, where the product is present at less than 10% concentration and is less volatile than water. Identification of sorbents, for example, that remove the solutes from the water could be an important first step in improving the process.

6.1.9 Agricultural Chemicals and Food

Bioprocess engineering in agriculture and the food industry involves the application of biocatalysts (living cells or their components) to produce useful and value-added products, and it offers opportunities to design and produce new or improved agricultural and food products and their manufacturing processes. This will likely have a great impact on the U.S. food-processing industry, which has estimated annual sales of $255 billion. In our increasingly health-conscious society, genetically engineered microorganisms and specialty enzymes will find increased use in improving the nutritional, flavoring, and storage characteristics and safety of food products. Products under development range from genetically improved strains of freeze-resistant yeast used in frozen bakery products to phage-resistant dairy (yogurt) starter cultures. Chymosin, a product of recombinant *E. coli*, is already used in the milk-clotting step of cheese manufacture, and a recombinant maltogenic amylase is being used as an antistaling agent. Enzyme-based immunoassays could develop into a widely used method for detecting pesticides in foods at parts-per-billion concentrations. Challenges that must be addressed include the economics of production and regulatory issues (Glaser and Dutton, 1992).

The most important applications of bioprocess-engineering research and development related to agriculture and food involve production of agricultural chemicals for control of animal and plant diseases, growth-stimulating agents for improved yield, and biological insecticides and herbicides; increasing bioprocess efficiencies for fermented foods, natural food additives, food enzymes as processing aids, and separation and purification of the products; use of plant-cell culture systems to produce secondary metabolites or chemical substances of economic importance; and efficient use of renewable biomass resources for production of liquid fuel and chemical feedstocks and efficient treatment and management of agricultural wastes and wastes from food-processing industries.

6.1.10 Plant-Cell Culture

The commercial potential of plant-cell tissue culture has not yet been fully recognized and is underexploited. Plant-cell tissue culture has two primary products: plant tissue for efficient micropropagation of plants and the use of plant-tissue culture to produce specialty chemicals.

Plant-cell, -tissue, and -organ cultures can be used in processes analogous to traditional fermentation processes for producing chemicals. Although less than 5% of the world's plants have even been identified taxonomically, from among the known plants over 20,000 chemicals are produced—about 4 times as many as from all microorganisms. Very few of the chemicals in pure or semipure form have been tested for their pharmacological activity for other uses. The enzymatic systems in plants can be used to generate completely new compounds when supplied with analogues of natural substrates; thus, plants contain an underused biochemical diversity. Even the limited use of this vast biochemical potential has had important impacts on mankind; in western countries, about one-fourth of all medicines are derived from compounds extracted from plants. Other plant products are used as flavors, fragrances, or pesticides.

Plant-cell tissue culture to produce chemicals commercially has been exploited in Japan, although regulatory approval for medicinal uses has proved difficult and commercial production is restricted to food uses and pigment production. In Japan, a government-sponsored consortium of universities and corporations was recently developed to establish a foundation for plant-cell culture exploitation (i.e., a precompetitive research thrust). In the United States, plant-cell tissue is not being exploited for chemical production, although at least two companies are actively developing processes for the production of the chemotherapeutic agent taxol.

The major technical barriers to the commercial exploitation of plant-cell tissue culture are low growth rates and relatively low product yields. To mitigate those problems, research is needed in subjects as diverse as bioreactor strategies to maintain high-density cultures and enable large-scale production of chemicals through organ cultures and a mechanistic understanding of the role of elicitors in activating pathways for secondary metabolites that could lead to higher productivities of compounds with therapeutic value.

6.1.11 Plants and Seeds

Basic research in plant molecular biology has allowed the clonal production of plants through the process of somatic embryogenesis, in which somatic cells develop through the stages of embryogenesis to yield whole plants without gamete fusion. Somatic embryos have been induced from a variety of plant tissues, and this system is commercially attractive for the high-volume multiplication of genetically improved embryos in culture. The clonal embryos are synthetic "seeds" that can be delivered to commercial growers. For many applications, somatic embryos have powerful advantages over conventional clonal propagation methods and other in vitro regeneration systems for mass propagation. One advantage is the very high multiplication rates. Depending on the plant species, virtually unlimited numbers of embryos can be generated from a single explant. A second advantage is that, for many species, growth and tissue development of somatic embryos can be carried out in a liquid medium. That fact gives somatic embryogenesis the potential to be combined with engineering technology to create large-scale culture systems. The development of such engineering technology is the limiting step in commercialization.

The use of submerged liquid culture for the efficient mass production of embryos, artificial seeds, or plant propagates is a promising industrial technique now used in Israel and other countries. For crop and forest plants, micropropagation on solid medium is too expensive. The use of bioreactors, which reduces labor costs greatly, will probably be necessary for mass production of crop and forest plants generated by genetic engineering or nontraditional breeding methods. The key engineering problem in such systems is the control of environmental conditions necessary for development of organized tissues from unorganized, minimally differentiated tissues. Fundamental studies on the interaction of concentration gradients with cellular developmental processes are required.

6.1.12 The Environment

Many environmental problems will require bioprocess engineering. Some are discussed elsewhere in this report. Here we mention that, in addition to the types of problems commonly considered (minimization of manufacturing waste, treatment of municipal waste, environmentally friendly manufacturing), new classes of problems might arise from the concept of "life-cycle environmental responsibility." That idea is being actively considered in various European countries for such products as automobiles. The manufacturer would be responsible for its product throughout its life cycle, including its ultimate disposal. If that responsibility becomes a reality in any of the major markets, it will be necessary to develop new classes of processes for final disposal of components. The need for biodegradable polymers, for example, coupled with efficient biodegradation systems for disposing of manufactured components, would become a reality. Large volume and low cost in operation would be essential. Because there are virtually no counterparts for them at present, there is no experience to guide future development; these types of engineering and process problems would be truly novel.

The removal and oxidation of organic gases from contaminated air by microorganisms fixed in beds of soil, compost, or other solid materials might gain expanded acceptance as a process by which air is cleaned through biological means. Biofiltration has already enjoyed industrial success in Europe and Japan. Nonetheless, significant bioprocess engineering challenges remain in the use of organisms to remove gas-phase organic chemicals. These include gas-solid contacting, maintaining stable microbial populations, and predicting performance for scaleup purposes.

6.1.13 Space

The unique problems presented by the manned exploration and colonization of space pose major and important challenges for bioprocess engineering. Space biology—the behavior of living systems in a zero-gravity (and perhaps high-radiation) environment—is just beginning to be investigated. The long-term influence of the space environment on cellular development remains to be determined. In principle, however, a plausible (from an economic point of view) application of processing in space would be to produce new cell lines. *If* it is possible to carry out manipulations of genetic materials or cell types in space that cannot be conducted on earth and *if* the modifications of cell behavior or germ-line composition that result from the manipulations can be preserved on return to earth, very high value could, in principle, be achieved. Because cells can be propagated and relatively small volumes of starting material (genetically altered cells) can be converted into large numbers of product cells after return to earth, the very high cost of manipulations in space would have a smaller affect on the overall cost of genetic manipulation of cell lines than of products that are sold by weight. The development of reactor systems and assay systems that are appropriate for use in space thus represents an important investment in this speculative field.

Reduced gravity (Moon and Mars) and microgravity (space station) would present some unique opportunities for the study of complex biological systems. One new interest is in the possible use of a microgravity environment to produce three-dimensional differentiated tissue structures in the fluid phase without a solid support. That could lead to a new tool for tissue engineering and the study of cell differentiation and developmental biology. Limitations in the knowledge of how cell suspensions and cell-to-cell communication affect the growth of such biomaterials might be quickly overcome once experiments are carried out on a space platform in low earth orbit. The objective would be research on manufacturing technologies, rather than manufacturing itself. Specific objectives include developing and testing experimental models for testing mammalian cell and tissue properties and for providing the necessary nutrients, biomodifiers, gases, and control. Important studies in developmental biology should be possible, including the generation of high-order tissue morphology of primary cells and ultimately perhaps complete organ generation. Once developed, the necessary techniques could be brought back to earth for manufacture. Bioprocess engineering is an element of all phases of such a project, from the design and implementation of orbiting bioprocess laboratory experiments to the implementation of the results for manufacturing processes on earth.

The problem of development of life-support systems for humans and other organisms in space is separate and highly important (if specialized).

The engineering of even conventional equipment for the space environment presents a unique set of problems. If the National Aeronautics and Space Administration (NASA) continues with its long-term goal of manned exploration of space, a major component of the reduction of space biology to engineering systems appropriate for life support of astronauts on long voyages will require the development of a specialized, light-weight, gravity-insensitive set of operations for handling and maintaining the appropriate environments for biological systems in space. The number of appropriately trained bioprocess engineers required for that effort would be substantial. NASA supports relatively small programs in space biotechnology through the Microgravity Science and Applications Division (MSAD) and the Life Sciences Division. One field in which the lack of density-gradient-driven convection could be important is protein nucleation and crystallization. Several active research projects are supported by MSAD (NASA, 1991). Another potential use of microgravity is in separation of large biological molecules (such as chromosomes) or cells (such as lymphocyte subpopulations), in which again gravity-induced sedimentation and density-gradient-driven convection could destroy separation ability.

This committee feels that space manufacturing of biological products for use on earth will not be fiscally feasible in the near future. For example, improvements in earth-based crystallization and separation systems and rapid developments in three-dimensional matrices for tissue engineering will make space-manufactured biomaterials too expensive. The space environment does provide a laboratory for interesting experiments in developmental biology and physiology, which should be pursued. Integration of those experiments into the space stations or the Moon- and Mars-based modules will require substantial collaboration between bioprocess engineers and basic biological scientists. That is particularly true because changes in local mass transport and fluid mechanics can often have important biological consequences (Nollert et al., 1991) on cell growth and structure formation, which must be separated from any effects of microgravity alone.

6.2 DEFENSE AND NATIONAL SECURITY

6.2.1 Cleanup

The Department of Defense (DOD) and Department of Energy face enormous problems in cleanup of military bases as they are closed or mothballed and in cleanup of weapons laboratories. In the former, biological methods might play an important part. Much of the contamination in military bases is in the form of hydrocarbons, explosives, and related chemicals that are intrinsically biodegradable. If it proves economically feasible to use biodegradation rather than, say, incineration or controlled detonation to dispose of unwanted ordnance and to clean up contaminated areas, the scale of the problem will require the development of new types of processes that can accommodate very large volume and low cost. Because military bases will be under government control, the sites could provide excellent opportunities for prototyping and developing cleanup technology that would be transferable directly into the civilian sector for municipal and hazardous-waste cleanup problems.

The weapons laboratories present special problems. The central problem is the management of radioisotopes, often in very dilute form as ground or water pollutants. Even here, however, biological systems might be useful. A key element in the cleanup of weapons laboratories is the reconcentration of dilute radioisotopes into concentrated form. Some microorganisms are very effective in mobilizing and concentrating specific elements. The development of appropriate organisms that would concentrate radioisotopes is not out of the question, but the engineering of systems required for processing contaminated materials and soils and for manipulating the biological products that would be produced does not now exist.

6.2.2 Chemical and Biological Warfare

A continuing problem in DOD concerns chemical and biological warfare. The stated U.S. position is to be prepared in a defensive mode, but not to be involved in production of biological weapons. The importance of either chemical or biological agents as practical weapons systems remains to be established. They are, however, systems for which protection must be available, and the potential for the use of either chemical or biological agents in the context of terrorism remains important. The defensive requirement for the ability to produce vaccines in large quantities with high flexibility for protection of military personnel (and, in some circumstances, civilian personnel) remains important. Some materials, like abzymes, have been considered for protection against chemical threats. Although that type of vaccine and countermeasure production can be carried out with conventional technologies, the opportunity to increase the responsiveness of the manufacturing systems, to allow manufacturing to produce tailored agents rapidly to meet new threats, and to lower the cost of production, particularly in episodic high-volume use, will benefit from development of new types of biological processes and manufacturing systems.

6.2.3 Stabilization of Developing Countries

A crucial element in national security is the stabilization of developing countries. Those countries might suffer from chronic deficits in food, energy, and transportation. They could, at the same time, have adequate sources of renewable resources and abundant and inexpensive labor. A plausible connection between needs and available resources lies in the production of foods and energy from renewable resources with local labor and local processing facilities. Although this type of research is not the kind that is best done in the United States, it represents a worthwhile investment for the United States, through such international organizations as the World Health Organization, as a way of improving the standard of living and political stability in developing countries.

6.3 NEEDS

To meet the challenges posed by the long-term opportunities in biological engineering and manufacturing, the field of bioprocess engineering must address education and training, technology, and manufacturing.

6.3.1 Education and Training

In advanced bioprocesses (as in the ones that are of current interest), the key issue is education and training. For the United States to take maximal advantage of opportunities in biology, it will be necessary to have engineers who are thoroughly familiar with current biology and biologists who are interested in engineering. The development of programs to couple training in engineering with training in molecular biology and other life sciences is of the highest priority.

6.3.2 Technology

The development of analytical systems, sensors, and methods for production of ultrapure products is crucial in all manufacturing processes (especially for therapeutic products) based on molecules. Those systems will be as important for nucleic acids, nucleic acid analogues, and carbohydrates as for proteins—or perhaps more important. The details of the systems for nucleic acids, carbohydrates, and proteins will, however, each be substantially different. Bioprocess engineering should begin to develop appropriate sensors and purification systems for the next generation of products.

A number of applications in human therapy will require the isolation and clonal amplification of specific cell types (perhaps with stimulation by specific growth factors, cytokines, or antigens). Cell sorting is a laboratory technique. The engineering development of processes capable of inexpensive, automated, large-volume cell sorting will be important for projects that rest heavily on mammalian cells (as opposed to molecules).

Bioprocess engineering went through a phase in which substantial work was invested in sophisticated reactors to obtain protein products from various types of cells. In general, those reactors have not been used: it has proved simpler to use relatively conventional reactors for such manufacturing processes. For more complex systems, however—especially of the sort required for organ culture and culture of other differentiated structures—new types of reactors will almost certainly be required.

6.3.3 Manufacturing

Manufacturing is the ultimate product of engineering. Experience in the more classical kinds of manufacturing—automobiles, machine tools, electronic devices, and consumer appliances—indicates that a key to long-term profitability and quality is the integration of design, engineering, and manufacturing. Such phrases as "design for manufacturing," "design for assembly," and now "design for disassembly" are common in industries involved in mechanical and electronic systems. They are less common in biomanufacturing. The development of disciplines that use engineering to couple early-stage science and prototyping with efficient manufacturing will be important in bioprocess engineering. In addition to "design for production," there might also be a discipline of "design for regulatory clearance"–the design of processes that are intrinsically robust to process variation and that will convincingly be presentable to the Food and Drug Administration and other regulatory agencies as safe and sure to yield products with acceptably small degrees of variance.

6.4 RECOMMENDATIONS

The outstanding performance of the United States in the basic life sciences should be maintained. The discoveries emanating from the basic life sciences provide the fundamental knowledge from which new concepts for products and biologically based manufacturing systems are derived. The committee strongly recommends that federal funding of biotechnology research be extended to support efforts that provide the science and technology base for producing and manufacturing products from biology. Targeted long-term research support would speed the development of commercial products, provide the trained personnel needed to support industrial activities, protect the entry-level U.S. products, provide the basis for low-cost production of the largest-volume (and highest-revenue) products, and help to integrate processes and concepts from biological science and bioprocess engineering. The United States has made an enormous, and enormously successful, investment in basic biological science. To protect that investment, and to capitalize on it, there must be an investment in bioprocess engineering.

6.5 REFERENCES

Glaser, V., and G. Dutton. 1992. Food processors seek to adapt bioproducts for large-scale manufacturing. GEN 12(2):6-8.

Hein, M. B., Y. Tang, D. A. McLeod, K. D. Janda, and A. Hiatt. 1991. Evaluation of immunoglobulins from plant cells Biotechnol. Prog. 7:455-461.

Hiatt, A., R. Cafferkey, and K. Bowdish. 1989. Production of antibodies in transgenic plants. Nature 342:76-78.

Luckow, V. A. 1990. Cloning and expression of heterologous genes in insect cells with baculovirus vectors. In Recombinant DNA Technology and Applications, C. Ho, A. Prokop, and R. Bajpai, eds. New York: McGraw-Hill Book Company.

Nagasawa, T., and H. Yamada. 1989. Microbial transformations of nitriles. Trends Biotechnol. 7(6):153-158.

NASA (National Aeronautics and Space Administration). 1991. Pp. 87-104 in Microgravity Science and Application Program Tasks, Technical Memorandum 4284, 1990 Revision. Washington, D.C.: National Aeronautics and Space Administration.

Nollert, M. U., S. L. Diamond, and L. V. McIntire. Hydrodynamic shear stress and mass transport modulation of endothelial cell metabolism. Biotechnol. Bioeng. 38:588-602.

Shuler, M. L., T. Cho, T. Wickham, O. Ogonah, M. Kool, D. A. Hammer, R. R. Granados, and H. A. Wood. 1990. Bioreactor development for production of viral pesticides or heterologous proteins in insect cell cultures. Ann. NY Acad. Sci. 589:399-422.

Bibliography

The reports listed here provide useful background and further details on bioprocess engineering, biotechnology, and policy issues.

Burrill, G. S., and K. B. Lee, Jr. 1991. Biotech '92: Promise to Reality, An Industry Annual Report. San Francisco: Ernst & Young.

Council on Competitiveness, Office of the Vice President. 1991. Report on National Biotechnology Policy. Washington, D.C.

Federal Coordinating Council for Science, Engineering, and Technology. 1992. Biotechnology for the 21st Century. Washington, D.C.: U.S. Government Printing Office.

Global Competitiveness Corporation and Technology International, Inc. 1991. Survey of Direct U.S. Private Capital Investment in Research and Development Facilities in Japan. Final Report for Science and Engineering Indicators Program, Grant No. SRS-8912547. Washington, D.C.: National Science Foundation.

Good, M. L., J. K. Barton, R. Baum, and I. Peterson, eds. 1988. Biotechnology and Materials Science—Chemistry for the Future. Washington, D.C.: American Chemical Society.

Japanese Technology Evaluation Center. 1992. Bioprocess Engineering in Japan. NTIS Report PB92-100213, Washington, D.C.

Lanks, K. W. 1990. Academic Environment: A Handbook for Evaluating Faculty Employment Opportunities. New York: Faculty Press.

National Academy of Engineering, Committee on Time Horizons and Technology Investments. 1992. Time Horizons and Technology Investments. Washington, D.C.: National Academy Press.

National Academy of Sciences, Briefing Panel on Chemical and Process Engineering for Biotechnology. 1984. Research Briefings. Washington, D.C.: National Academy Press.

National Research Council, Committee on a National Strategy for Biotechnology in Agriculture. 1987. Agricultural Biotechnology: Strategies for National Competitiveness. Washington, D.C.: National Academy Press.

National Research Council, Committee on Bioprocessing for the Energy-Efficient Production of Chemicals. 1986. Bioprocessing for the Energy-Efficient Production of Chemicals. Publication NMAB 428. Washington, D.C.: National Academy Press.

National Research Council, Committee on Chemical Engineering. 1988. Frontiers in Chemical Engineering: Research Needs and Opportunities. Washington, D.C.: National Academy Press.

National Research Council, Engineering Research Board. 1987. Directions in Engineering Research: An Assessment of Opportunities and Needs. Washington, D.C.: National Academy Press.

National Science Foundation. 1987. Biotechnology Research and Development Activities in Industry: 1984 and 1985. Surveys of Science Resources Series, Special Report NSF 87-311. Washington, D.C.: National Science Foundation.

National Science Foundation. 1990. Engineering Research Centers: A Partnership for Competitiveness. Washington, D.C.: National Science Foundation.

Office of Technology Assessment, U.S. Congress. 1991. Biotechnology in a Global Economy. B. Brown, ed. Report No. OTA-BA-494. Washington, D.C.: U.S. Government Printing Office.

Olson, S. 1986. Biotechnology: An Industry Comes of Age. Academy Industry Program of the National Academy of Sciences, National Academy of Engineering, and Institute of Medicine. Washington, D.C.: National Academy Press.

U.S. Department of Agriculture, New Farm and Forest Products Task Force. 1987. New Farm and Forest Products: Responses to the Challenges and Opportunities Facing American Agriculture. Washington, D.C.: U.S. Department of Agriculture.

U.S. Department of Agriculture, Office of Agricultural Biology. 1990. Biotechnology at USDA. Washington, D.C.: U.S. Department of Agriculture.

U.S. Department of Energy, Office of Health and Environmental Research. 1990. Form and Function: Perspectives on Structural Biology and Resources for the Future. Washington, D.C.: U.S. Department of Energy.

Appendix A

Biographical Sketches of Committee Members

Michael R. Ladisch is professor of bioprocess and agricultural engineering and group leader of the Research and Process Engineering Group in the Laboratory of Renewable Resources Engineering at Purdue University. He received his B.S. degree in chemical engineering from Drexel University in 1973 and M.S. and Ph.D. degrees in chemical engineering from Purdue University in 1974 and 1977, respectively. Dr. Ladisch's research interests are in bioseparations, kinetics of biochemical reactions, chemical-reaction engineering, and biomass conversion. In 1978, he joined the faculty at Purdue University as assistant professor, and he has been a full professor since 1985. He has written numerous papers and book chapters and has worked actively with industry in implementing fundamental research results in the form of new process technology. He received the U.S. Presidential Young Investigator Award in 1984 and the James Van Lanen Distinguished Service Award of the American Chemical Society's Biochemical Technology Division in 1990.

Charles L. Cooney is professor of chemical engineering and biochemical engineering in the Department of Chemical Engineering, co-director of the program on the pharmaceutical industry, and associate director for industrial activities at the Biotechnology Processing Engineering Center at Massachusetts Institute of Technology, Cambridge, Massachusetts. He obtained his bachelor's degree in chemical engineering from the University of Pennsylvania in 1966 and his master's and Ph.D. degrees in biochemical engineering from MIT in 1967 and 1970, respectively. After working briefly at the Squibb Institute for Medical Research, he joined the faculty of MIT as an assistant professor in 1970 and has been a full professor since 1982. His research interests are in computer control of biological processes, downstream processing for recovery of biological products, bioreactor design and operation, and manufacturing strategies in the pharmaceutical industry. He has received the Institute of Biotechnological Studies 1989 Gold Medal, the Food, Pharmaceutical and Bioengineering Award from the American Institute of Chemical Engineers, and the James Van Lanen Distinguished Service Award from the American Chemical Society's Division of Microbial and Biochemical Technology and was recently elected to the American Institute of Medical and Biochemical Engineers. He serves as a consultant to or director of a number of biotechnology and pharmaceutical companies and is on boards of several professional journals.

Robert C. Dean, Jr. is the founder of two bioprocessing companies: Verax (production mammalian-cell culture systems) and Synosys (now PerSeptive Biosystems, Inc., production protein-purification systems). He has also founded four other companies: Creare, Inc., Hypertherm, Inc., Creare Innovations, Inc. (now Spectra, Inc.), and Dean Technology, Inc. He holds Sc.D. (1954), M.S. (1949), and B.S. (1948) degrees in mechanical engineering from the Massachusetts Institute of Technology. In 1987, Dr. Dean founded the American Society of Mechanical Engineers' Bioprocess Engineering Program. Dr. Dean was an assistant professor of mechanical engineering at MIT in 1951-1956. He was head of advanced engineering at Ingersoll-Rand Company in 1956-1960. He became associate professor in 1961 and later professor of engineering at Dartmouth College. He is a professor of engineering (adjunct) at Dartmouth and at Northeastern University. He heads his firm, Dean Technology, Inc., Lebanon, New Hampshire, where he is developing innovation processes, materials, and equipment for the manufacture of biopharmaceuticals specifically and biochemicals in general; for medical applications; for machining difficult materials; and for advanced materials fabrication. He is a member of the National Academy of Engineering.

Arthur E. Humphrey received his B.S. and M.S. degrees in chemical engineering from the University of Idaho in 1948 and 1950, respectively. He received a Ph.D. in biochemical engineering from Columbia University in 1953. Dr. Humphrey joined the University of Pennsylvania in 1953 and conducted research and taught there until 1980. In 1960, he obtained an M.S. degree in food technology from the Massachusetts Institute of Technology. At the University of Pennsylvania, Dr. Humphrey served as chair of the Chemical Engineering Department for 10 years and dean of engineering and applied science for 8 years. He moved to Lehigh University in

1980, where he served for 6 years as provost and academic vice president and then held the positions of T. L. Diamond Professor of Biochemical Engineering and director of the Center for Molecular Bioscience and Biotechnology. In 1992, he became director of the Biotechnology Institute and professor of chemical engineering at Pennsylvania State University. Dr. Humphrey's work on the design and control of bioprocesses has yielded more than 250 research papers, three books, and four patents. He is a fellow of the American Institute of Chemical Engineers. In 1984, Dr. Humphrey chaired the Research Briefing Panel for the Office of Science and Technology Policy on "Chemical and Process Engineering for Biotechnology." He is a member of the National Academy of Engineering.

T. Kent Kirk received his B.S. degree from Louisiana Polytechnic University in 1962 and his M.S. and Ph.D. (biochemistry, plant pathology) from North Carolina State University in 1964 and 1968, respectively. He completed postdoctoral studies in organic chemistry at Chalmers University of Technology in Göteborg, Sweden. Dr. Kirk is now director of the Institute for Microbial and Biochemical Technology, USDA Forest Products Laboratory, Madison, Wisconsin. He is also professor in the Department of Bacteriology, University of Wisconsin, Madison, and adjunct professor, Department of Wood and Paper Science, at North Carolina State University. Dr. Kirk's research has focused almost entirely on the microbiology, chemistry, and biochemistry of the fungal degradation of lignin and on industrial applications of fungi. He has received a number of awards and honors, including the USDA Superior Service Award and the Marcus Wallenberg Prize from Sweden. In 1988, he was elected to the U.S. National Academy of Sciences.

Larry V. McIntire is the E. D. Butcher Professor of Chemical and Biomedical Engineering at Rice University. He is also chair of the Institute of Biosciences and Bioengineering and director of the Cox Laboratory for Biomedical Engineering within the institute. Dr. McIntire received his B.Ch.E. and M.S. degrees from Cornell University in 1966 and his Ph.D. from Princeton University in 1970—all in chemical engineering. He has been at Rice University since 1970. His research interests include the effects of flow on mammalian-cell metabolism, molecular mechanisms of cell adhesion, tissue engineering, mammalian-cell culture, and bioengineering aspects of vascular biology. Dr. McIntire is the recipient of a National Institutes of Health MERIT Award and is a founding fellow of the American Institute of Medical and Biological Engineering. Dr. McIntire was the 1992 ALZA Distinguished Lecturer of the Biomedical Engineering Society and the 1992 recipient of the American Institute of Chemical Engineering Food, Pharmaceutical, and Bioengineering Division Award.

Alan S. Michaels received the Sc.D. degree from the Massachusetts Institute of Technology in chemical engineering in 1948, joined the faculty of MIT on graduation, and became full professor in 1961. In 1962, he founded and became president of Amicon Corporation, which pioneered the development of membrane ultrafiltration as a novel molecular-separation process. Dr. Michaels resigned his tenure position at MIT in 1966 to devote full time to the management of Amicon, where he remained as president until 1970, when he founded and became president of Pharmetrics, Inc., Palo Alto, California, a company engaged in research and development of controlled drug-delivery systems, in collaboration with ALZA Corporation. Pharmetrics was merged into ALZA in 1972, and Dr. Michaels became senior vice president and technical director of ALZA and president of its research division. He was instrumental in the development of ALZA's transdermal (TRANSDERM) and oral osmotic (OROS) delivery systems, which are now in widespread use around the world. In 1976, he became professor of chemical engineering and medicine at Stanford University, where he remained until 1981. He then returned to New York and Boston to conduct an independent industrial-consulting practice. In 1986, he joined the chemical-engineering faculty at North Carolina State University as distinguished university professor. He retired from his faculty position in 1989 (he is currently distinguished professor, emeritus) and returned to Boston to resume his full-time industrial-consulting practice as president of Alan Sherman Michaels, Sc.D., Inc. He is the author or coauthor of more than 140 refereed journal articles and contributed to eight textbooks and monographs. His honors include the McGraw-Hill Outstanding Personal Achievement Award in Chemical Engineering; 37th Institute Lecturer of the American Institute of Chemical Engineers (AIChE); the Food, Pharmaceutical, and Bioengineering Award of AIChE; the Materials Engineering and Science Award of AIChE; and the ACS Award in Separation Science and Technology. He was honored as Ninth Centennial Chemical Engineering Lecturer, University of Bologna, Bologna, Italy, in 1988. He is a member of the National Academy of Engineering.

Paula Myers-Keith is director of bioprocess research at Pitman-Moore, Inc., where she is responsible for directing research programs in molecular biology, microbial genetics, fermentation-process development, and biosep-arations. In addition, she chairs Pitman-Moore's corporate technology-assessment group with responsibility for evaluation of new technology and its application to novel animal products or veterinary pharmaceuticals. Dr. Keith holds a bachelor's degree in biology (1971, West Liberty State College), an M.S. in microbiology (1973, West Virginia University), and a Ph.D. in microbiology (1978, Virginia Polytechnic Institute). After a year of postdoctoral research, she joined Pitman-Moore, where her primary research interest has been fermentation-process development of new animal products (recombinant growth hormones, polyether antibiotics, vaccines, anabolics). She has written more than 25 papers and patents. Dr. Keith has served on the board of directors of the Society for Industrial Microbiology for several years and recently completed a term as president. She is a member of the editorial board of the *Journal of Industrial Microbiology* and was the 1992 recipient of the Charles Porter Award.

Dewey D. Y. Ryu is professor in the Department of Chemical Engineering and director of the Biochemical Engineering Program at the University of California, Davis. He received his B.S. degree in chemical engineering in 1961 and Ph.D. degree in biochemical engineering in 1967—both from the Massachusetts Institute of Technology. He worked for several years as a senior research engineer at the Squibb Institute for Medical Research (now Bristol-Myers Squibb). He was one of the founding members of the Korea Advanced Institute of Science in Seoul, where he was professor and chair of the Department of Biotechnology. He joined the University of California, Davis in 1981. His research interests include recombinant and nonrecombinant fermentation technology, biocatalysis and enzyme engineering, large-scale mammalian- and plant-cell cultivation technology for production of biologically active compounds of medical importance, use of renewable resources, and bioseparations. He has contributed about 160 scientific publications and 17 invention patents dealing with novel products, bioprocesses, and bioreactor design, all applicable to production of new gene products, antibiotics, steroid hormones, enzymes, and food additives. He has provided a wide range of professional services to industry, national and international professional organizations, government organizations, and other academic institutions, including membership of the editorial boards of biotechnology journals and of study groups and review committees for the National Institutes of Health, the National Science Foundation, the National Research Council, the United Nations, and other academic and research organizations.

James R. Swartz obtained his B.Ch.E. degree from the South Dakota School of Mines and Technology. After working for 2 years for Union Oil Co. of California, he attended the Massachusetts Institute of Technology, where he earned his M.S. and D.Sc. in chemical engineering and biochemical engineering, respectively. His focus on the development and control of fermentation processes led him to a scientific exchange visit to the USSR and to employment at Eli Lilly and Co. in Indianapolis. In 1981, he went to Genentech, first serving as a scientist and then forming and serving as director of the Department of Fermentation Research and Process Development. He returned to active research and development, focusing on the expression and secretion of mammalian proteins from *E. coli* in 1988. He is also active in the development of large-scale bacterial fermentation processes and is serving as insulin-like growth factor-I project team leader at Genentech. He is a member of several professional societies and has served as program chairman and division chairman for the American Chemical Society Division of Biochemical Technology.

Daniel I. C. Wang is Chevron Professor of Chemical Engineering and director of the Biotechnology Process Engineering Center at the Massachusetts Institute of Technology. Professor Wang received his B.S. degree in chemical engineering and his M.S. degree in biochemical engineering from MIT in 1959 and 1961, respectively. He received his Ph.D. in chemical engineering from the University of Pennsylvania in 1963. Professor Wang's research interests include transport phenomena in animal-cell bioreactors, biosensors in bioprocess monitoring and control, protein purification and protein refolding in downstream processing, bioreactor design in viscous fermentations, and oxygen transfer in fermentation vessels. His work has produced four books, more than 150 publications, and 11 patents. He currently serves on the National Institutes of Health Board on Biotechnology Policy, the National Research Council Board on Chemical Sciences and Technology and Board on Biology, the NRC Committee on Biotechnology, the National Academy of Engineering Peer Review Committee, the Republic of China Biotechnology Center Advisory Board, and the Singapore Science Council Advisory Board. Professor Wang was elected to the National Academy of Engineering in 1986.

Janet Westpheling is an assistant professor of genetics at the University of Georgia. Dr. Westpheling received her B.S. degree in microbiology from Purdue University and her Ph.D. in genetics from the John Innes Institute in 1980 and was a postdoctoral fellow at Harvard University. Her primary research interest involves the control of gene expression in *Streptomyces,* with emphasis on the study of carbon utilization and primary metabolism and the strategies used by bacteria to regulate genes involved in morphogenesis and secondary metabolism.

George M. Whitesides received his A.B. degree from Harvard University in 1960 and his Ph.D. from the California Institute of Technology in 1964. He was a member of the faculty of the Massachusetts Institute of Technology from 1963 to 1982. He joined the Department of Chemistry of Harvard University in 1982 and was department chairman in 1986-1989. He is now Mallinckrodt Professor of Chemistry at Harvard University. His present research interests include biochemistry, surface chemistry, materials science, reaction mechanisms, and catalysis. His recent advisory positions include service on the National Research Council Board on Chemical Sciences and Technology (1984-1989), the Defense Advanced Research Projects Agency Defense Science Research Council (1984-), the MIT Advisory Committee for Lincoln Laboratory (1985-), the NRC Naval Studies Board (1989-), the NRC Board on Science, Technology, and Economic Policy (1991-), and the National Science Foundation Materials Research Advisory Committee. He received an Alfred P. Sloan Fellowship in 1968, the American Chemical Society Award in Pure Chemistry in 1975, the Harrison Howe Award (Rochester Section of the American Chemical Society) in 1979, an Alumni Distinguished Service Award (California Institute of Technology) in 1980, the Remsen Award (American Chemical Society, Maryland) in 1983, and an Arthur C. Cope Scholar Award (American Chemical Society) in 1989. He is a fellow of the American Association for the Advancement of Science and a member of the American Academy of Arts and Sciences and the National Academy of Sciences.

Appendix B

Invited Speakers at
Committee Meetings

Stanley Abramowitz, National Institute of Standards and Technology
Maurice M. Averner, National Aeronautics and Space Administration
Duane Bruley, American Society of Mechanical Engineers
Joan Burrelli, American Chemical Society
Brad Carpenter, National Aeronautics and Space Administration
Marvin Cassman, National Institute of General Medical Sciences, National
    Institutes of Health
Mark Dibner, North Carolina Biotechnology Center
Alan Fechter, National Research Council
Robert Frederick, Environmental Protection Agency
Bruce Guile, National Academy of Engineering
Fred Heineken, National Science Foundation
Daphne Kamely, SARD-TR, Pentagon
Leonard Keay, Department of Energy
Barbara Lujan, National Aeronautics and Space Administration
Hiram Larew, Agency for International Development
Marshall Lih, National Science Foundation
Stephen A. Lingle, Environmental Protection Agency
Richard F. Moorer, Department of Energy
Robert Parry, Department of Agriculture
Lura Powell, National Institute of Standards and Technology
Lynn Preston, National Science Foundation
Linda Schilling, Department of Energy
Paul Scott, Department of Energy
Carolyn Shettle, National Science Foundation

James J. Valdes, U. S. Army
Ruxton Villet, Department of Agriculture
Ron White, National Aeronautics and Space Administration
Daniel E. Wiley, Department of Energy

Index